Foundation of Environmental Science

Foundation of Environmental Science

Abhijit Mitra, Ph.D.
Department of Marine Science
University of Calcutta
56, Belgachia Road
CALCUTTA-700 037

Narendra Publishing House
Delhi

First Published in 2000

ISBN 81-85375-62-3 HB
ISBN 81-85375-63-1 PB

Published by:

Narendra Publishing House
PUBLISHER AND DISTRIBUTORS
1417, Kishan Dutt Street, Maliwara,
DELHI-110 006 (India) ✆ 3268470, 3259412

Printed in India

DEDICATION

This book is dedicated to Mother Earth, who sustains us and all other organisms, to my parents who showed me the light of life and to my career guide, Prof. Amalesh Choudhury for his affection, support and inspiration.

PREFACE

The new millennium has stepped with new hopes of conservation, rational exploitation and systematic management of natural resources, but pollution, population and poverty are the three adverse forces to nip these hopes in the bud. In the student community, there is an unfortunate gap in understanding of different ecological terminologies and concepts, although they should be the pioneer to give this earth a proper environment, where golden sun will shine, the birds will chirp in the green belt and the reflection of blue sky will be felt in the crystal clear water of the oceans.

Although the task of building a sound knowledge on the dynamics of ecosystem is difficult, but generation of basic concept on ecology and environment can be done through some common incidences around us in our day-to-day life. The sudden awakening of volcanoes from their dormant state in different volcanic belts of the earth, the occurrence of devastating flood, the generation of heat wave, the phenomenon of super cyclone or El-Nino are not the stray events - rather they are all linked with the principles of ecosystem, which we are violating in every steps with our high technology and scientific machinery. The net result is an adverse thrust from the nature due to which lives and properties of millions of people come to an end or gradually deteriorate. Although it is too late, but today man has realized that he must live on the lap of nature and must not attempt to take nature in the grip. With this background, man is today adopting various ecofriendly steps for implementing any projects through which the ambient environment is least disturbed.

The present book aims to discuss the various diversities existing in the reservoir of nature and the need for conservation of this unique biodiversity. Most ecological terminologies have been explained citing the Indian example, so that the pulse of the subject can be easily felt by the students.

The book also highlights several bioremediation technologies which are not only ecofriendly, but are also related to conservation and preservation of our natural resources.

Since the environmental thinking and terminologies are ever increasing, the selected references are given at the end of the book are provided to facilitate the access and availability of relevant literature to the readers. All comments from users on omissions are shortfalls will be most welcome.

Abhijit Mitra

CONTENTS

Chapter 1:

ECOSYSTEM

1.1. DEFINITION AND CONCEPT

In 1866 German biologist Ernst Haeckel derived the term ecology from two Greek words: ***Oikos,*** meaning *house* or *dwelling place* and ***Logos,*** meaning *study of.* Infact the survival, growth or normal life processes of living organisms in their dwelling places are dependent on several biotic and abiotic parameters and ecology encompasses this particular branch of science. In more formal terms, ecology is the study of interactions among organisms and between organisms and the chemical and physical factors making up their environment. This study is usually carried out through an indepth cross-sectional examination of various ecosystems like forests, deserts, ponds, estuaries or oceans considering all their biotic components and abiotic environments.

An ecosystem, thus, consists of all the different types of organisms in an area and the physico-chemical variables of the ambient environment with which they interact. It may be defined as the self-regulating community of plants and animals interacting with one another and also with their nonliving environment. The term "*ecosystem*" was first coined by Tansley in the year 1935. All the populations of a given area cannot be separated functionally from the environment, rather both the community and the environment function together as a system what is known as *ecological system* or *ecosystem* (Whittaker, 1960; Odum, 1966). An ecosystem is described by the quantity of material and energy that is contained in each of its major compartments together with rates of flow of energy and materials among those compartments. The flow of energy is unidirectional, whereas the flow of materials is cyclic.

The concept of an ecosystem is essential for two reasons. Firstly, it helps to understand the modification of environment caused by many kinds of organisms which has profound influence on other types of organisms. Secondly it helps to understand the global biogeochemical cycles that support the foundation of life on earth and the various human activities that are strongly influencing these cycles. Today the concept of ecosystem finds wide application in different spheres of life like population control, flood control, town planning, resource conservation, pollution control, fisheries and Environmental Impact Assessment (EIA) that is necessary for initiating any industries, dams, bridges or highways.

1.2. COMPONENTS OF ECOSYSTEM AND THEIR INTER-RELATIONSHIP

All the ecosystems of the biosphere consist of various nonliving (abiotic) and living (biotic) components.

The non-living components of an ecosystem include various physical factors (like sunlight, humidity, atmospheric temperature, precipitation, water currents, waves, wind, etc.) and chemical factors (like the various elements or compounds in the atmosphere, hydrosphere or lithosphere that are essential for living organisms).

The different types of flora and fauna that constitute the living (biotic) components of an ecosystem are classified as producers, consumers and decomposers depending on the basis of their sources of energy and matter.

Producers are plants that can manufacture or synthesize their own food and encompass all the green plants of the earth. They can manufacture carbohydrates - glucose ($C_6H_{12}O_6$) from carbon-di-oxide (CO_2) and water (H_2O) utilising sunlight as the source of energy. Since producers are the only biotic component who can manufacture food, therefore all animals depend on them for their nourishment. Thus *all flesh on the earth is grass* is not an exaggerated statement. **Consumers** are organisms, mostly animals, that cannot manufacture or prepare their own food and depend directly or indirectly on producers for their nourishment. Depending on their food sources, consumers may be (1) primary consumers, which directly feed on plant like **herbivores** or (2) secondary consumers and higher level consumers, which feed on plant eating animals like the **carnivores. Decomposers** constitute another important group of biotic component, which obtain nutrients by breaking down complex molecules in the wastes and dead bodies of other organisms into simpler chemicals, some of which are returned to the soil and water for reuse by producers.

The inter-relationship between the components of an ecosystem can be visualized through food chains and food webs existing in the particular ecosystem. The food chain represents the sequence of transfer of energy in the form of food from organisms of one trophic level to organisms of another trophic level when one organism eats or decomposes another. In the language of Odum (1971) ".....food chain is the transfer of energy from the source in plants through a series of organisms with repeated stages of eating and being eaten.........". Food chains may be **(I)** Grazing food chain or **(II)** Detritus food chain.

(I) **Grazing food chain** starts from autotrophs i.e., producers (green plants) and then passes through herbivores (plant eating animals). Examples of grazing food chain is very common in forest ecosystem as shown below:

Grass	⇒	Rabbit	⇒	Fox	⇒	Tiger
(Producer)		(Primary consumer)		(Secondary consumer)		(Tertiary consumer)

(II) The term "*detritus*" is derived from the Latin word *deterere* meaning *wear away*. Actually detritus refers to all the particulate organic matter involved in the decomposition of dead organisms. It may also be defined as all types of biogenic material in various stages of microbial decomposition. The detritus, which is the outcome of microbial action, is derived from the breakdown of land plants, benthic algae and grasses, phytoplankton and zooplankton and is consumed by some small animals like insect larvae, copepods, nematodes etc. These detritus consumers are called **detrivores.** Detrivores are again consumed by smaller carnivores (small carnivorous fish), which in turn are preyed upon by larger carnivores (large fish) as shown below:

Decomposed organic matter
(Fallen leaves, decaying plants)

⇓

Detritus consumer
(Insect larvae, amphipods)

⇓

Small Fish

⇓

Large fish

Many workers have also classified food chain into **(1)** Predator chain, **(2)** Parasitic chain and **(3)** Saprophytic chain. The main criteria of difference between the predator chain and the parasitic chain is the size. In case of parasitic food chain the organisms at successive levels are smaller and smaller instead of being generally larger and larger. Thus roots of vegetable crops are parasitized by nematodes, which may be attacked by bacteria or other smaller organisms. Mammals and birds are commonly parasitized by fleas, which in turn have protozoan parasites of the genus *Leptomonas* indicating again a decrease in size with the increase of trophic level.

Food chains are not isolated sequences, but are interconnected with one another in all types of ecosystem. The interlocking pattern is usually referred to as *food web*. Thus food webs are complex network of many interconnected food chains and feeding interactions.

To understand the interplay of structure and function of biotic and abiotic components of an ecosystem, the entire phenomenon is considered from different angles. According to Evans (1956), Whittaker (1960), Odum (1966, 1971), the producer-consumer relationship highlights one kind of structure called the *trophic structure. Trophic level* is, therefore, regarded as steps or tiers through which food energy passes into an ecosystem. In grassland ecosystem, the autotrophic plants (producers) occupy the first trophic level, herbivores second trophic level, primary consumers third trophic level and secondary consumers fourth trophic level. Parasites also constitute further trophic levels as they form additional links to food chain.

Trophic structure and also trophic function may be represented by means of *ecological pyramids* in which the first trophic level or producer forms the base and successive levels, the consumers, make up the apex. Ecological pyramids may be broadly classified under three main headings, namely **(1)** the *pyramid of numbers* in which the number of individual organisms is depicted, **(2)** the *pyramid of biomass,* which is based on the total dry weight, caloric value or any other measure reflecting the total amount of living material, and **(3)** the *pyramid of energy,* which reflects the rate of energy flow through successive trophic levels considering the wastage of energy at each trophic level due to life processes like respiration.

Of the three types of ecological pyramids, the pyramid of energy gives by far the best overall picture of the functional nature of communities, since the number and weight of organisms that can be supported at any level in any situation depends not only on the amount of fixed energy present at any instant of time in the level just below, but also on the rate at which food is being produced. In contrast to the pyramids of numbers and biomass, which highlight pictures of standing states, i.e., organisms present at any particular instant of time, the pyramid of energy reflects the rates of passage of energy through food mass via the food chain. Its shape is not dependent on the variation of size of individuals and it also takes into consideration the wastage of energy during the process of energy transference through the trophic levels, which is in accordance with the second law of thermodynamics.

1.3. MAJOR TYPES OF ECOSYSTEM

1.3.1. Fresh-water ecosystem

Ecologists divide freshwater ecosystems into bodies of running water **(Lotic systems)** like rivers, springs, streams etc. and bodies of standing water **(Lentic systems)** like lakes, ponds etc.

Lotic system

The lotic systems encompass running water bodies with profuse dissolved oxygen in the aquatic phase. Rivers and streams which are unique examples of lotic systems, depend heavily on land, both for the water that fills their banks and materials that nourish the life within. Dead leaves and sewage from the adjacent towns and cities increase the nutrient level of the water bodies and sustain the existing food webs of the ecosystem. Nutrients are less abundant where water flows rapidly, because decaying organisms are quickly washed away. Dissolved oxygen in flowing water is usually abundant because air and water mix in turbulent areas. This creates a healthy environment for the growth and survival of the aquatic organisms. Dumping organic materials, in excess of the carrying capacity of the ecosystem, can deteriorate the water quality as their decomposition process uses the dissolved oxygen of the aquatic phase.

Different organisms are found in different parts of a river, each species adapted to local conditions. In swift current, some organisms hold onto any available substratum like rocks, logs etc. This is where to find algae, diatoms and mosses, as well as the snails that graze on them. Many insects at their larval and adult stages hold onto underside of rocks with hooks and suckers. The body contours of organisms in flowing water are usually streamlined for gliding through the current. As a river or a stream is fed by tributaries, it becomes larger and its currents slow. Such slower moving rivers and streams support more diverse life including crayfish, snails, bass, catfish etc. The muddy substratum at the bank offers a unique habitat for a variety of fauna like worms, molluscs etc.

Lentic system

Standing water bodies like lakes, ponds, bogs etc. are included under the lentic systems. Light penetrates to different degrees in the regions of a lake or a pond, and these differences determine the types of plants that live in a particular area. The **littoral zone** is the shallow region along the shore where light can penetrate to the bottom with sufficient intensity, so as to trigger the process of photosynthesis. Although some photosynthetic organisms in this region are free floating, most species are rooted to the bottom. Typical producers include cyanobacteria, green algae and diatoms as well as some macrophytes like water hyacinth, water lily etc. Animal life is also diverse, including damselflies, dragonfly nymphs, crayfish, rotifers, flatworms, hydra, snails, snakes, frogs and fishes.

The **limnetic zone** is the layer of open water that is penetrated by light. It is inhabited by phytoplankton, zooplankton and fish. The **profoundal zone** is the deep region beneath the limnetic zone where light does not penetrate. The biotic community in this zone depends on the rain of organic material from the surface layer and include mostly scavengers and decomposers such as bacteria, fungi and insect larvae.

Oxygen and mineral nutrients in the lakes are distributed unevenly. Concentration of dissolved oxygen is usually greater in the upper layers, where it comes from the atmosphere through the process of diffusion. The process of photosynthesis, being carried out in the upper layers of lakes and ponds by the green algae and macrophytes also increase the oxygen concentration. Deeper lakes in the temperate regions that have seasons often develop layers with very different water temperature. The layers prevent complete circulation in the lake. As a result of this **thermal stratification,** the redistribution of oxygen and nutrients occurs seasonally.

In the summer, the upper layer of a lake is heated by the sun, but the lower layers remain cold. Between these two layers is a third region, called the **thermocline,** where water temperature drops quickly. In the fall, the water temperature in the upper layer of the lake drops with the air temperature. Gradually the water temperature becomes the same throughout the lake. Wind then mixes the upper and lower layers, creating a **fall turnover** of nutrients and oxygen. During the winter, surface water becomes colder. When water cools to 4°C, the temperature at which it is most dense, it sinks. Water colder than this floats above the 4°C layer. With the warmth of the spring sun, when the temperature of the upper layer rises to 4°C, the water starts sinking, bringing oxygen with it. In turn, the nutrient rich lower layers circulate to the top. This **spring turnover** often results in great algal growth due to high nutrient concentrations at the upper layer called **blooms.**

On the basis of fertility and productivity, lakes are often classified into two types e.g., **(1)** ***oligotrophic*** **-** deep lakes, poor in nutrient level and productivity and **(2)** ***eutrophic*** **-** shallow lakes, rich in nutrients and productivity.

The biotic community in freshwater ecosystems like lakes and ponds encompasses different categories like **(a)** ***macrophytes*** **-** large plants, **(b)** ***planktons*** **-** small, free floating or drifting organisms that cannot move against the water current, **(c)** ***nektons*** **-** organisms that are able to swim at their will, **(d)** ***neustons*** **-** organisms (both resting and swimming types) that are confined to the water surface and **(e)** ***macrofauna*** **-** large animals that include the nektons and also benthic animals like snails, clams etc.

These organisms are intricately related to each other through food webs, in which phytoplanktons and macrophytes constitute the producer level, while zooplanktons, nektons, neustons and macrofauna are the consumers of different tiers.

1.3.2. Marine ecosystem

Covering 70% of the earth's surface and 11.2 kilometers deep in some zones, the ocean is the largest and most stable of biomes. Here, boundaries between habitats are classified by their proximity to the land. The marine ecosystem, thus encompasses the seashores as well as the ocean basins. This ecosystem has been classified by the ecologists from various angles like depth, type of bottom, light level etc.

Classification on the basis of substratum depth and submarine features

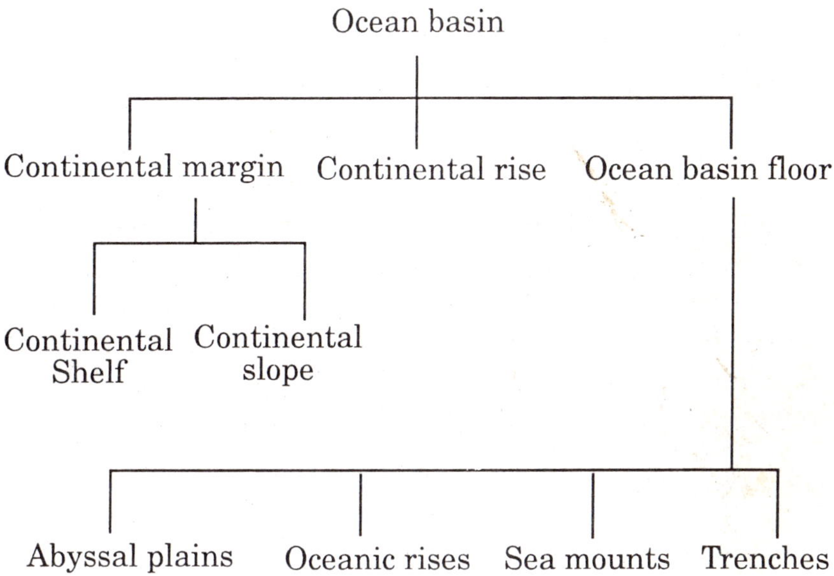

Continental margin

The continental margin includes the continental shelf and slope. The **continental shelves** fringe the continents and slope gently towards the ocean basins. They are relatively flat, generally less than 130 m deep. However, the range of the depth for the continental shelf is 30 m to about 275 m. The criterion for defining the seaward edge of the continental shelf is a marked change in slope. The continental shelf has a gradient of about 1:1000. Seaward from the continental shelf is the **continental slope** which extends to a depth of 1.6 to 3.2 Km. The continental slope has a gradient between 1:2 to 1:40. The slope is scarred by many submarine canyons which may be as deep as 1200 m. It is believed that these canyons have been cut by turbidity current of dense sediment-laden waters flowing down the surface of the continental slope.

Continental rise

Sediment from the continental shelf are carried into the deep ocean through the canyons present in the continental slope. At the foot of the continental slope, the suspended materials and sediments settle out, that slope gently towards the ocean floor. This region is called the continental rise and has a gradient between 1:50 to 1:800. The continental rise is fairly well developed in the Atlantic and Indian oceans, but is narrow or almost non-existent in the Pacific ocean.

Ocean basin floor

The ocean basin floor extends seaward from the continental rise or the continental margins and includes the abyssal plains, oceanic rises, sea mounts and trenches.

Abyssal plains are found at the base of the continental rise and are flat plains having gradient less than 1:1000. They are formed due to even deposition of sediments from the continental rise carried down by the turbidity currents. **Oceanic rises** refer to rather isolated areas which are elevated above the abyssal floor and are distributed sporadically on the ocean basin floor. They may vary from low hills to mountains as high as 1525 m. The example of Bermuda rise is very prominent in this case on which the

Bermuda Islands are formed. **Scamounts** are isolated peaks that rise several thousand meters above the sea floor. They are volcanic in origin and are found principally in the vicinity of the fault zones. **Trenches** are long narrow depressions in the ocean floor which are over 6100 m deep. The deepest known trench in the ocean compartment is the ***Mariana trench*** of the western North Pacific which is about 11,000 m deep. Trenches are invariably associated with the systems of active volcanoes, and are believed to be caused by down wrapping of the oceanic crust beneath the continental crust.

Classification of ocean water mass

The entire water mass above the ocean floor (benthic substratum) with all the diverse types of organisms is referred to as the pelagic zone and may be differentiated into neritic and oceanic zone. **Neritic zone** is the shallow water zone that extends from the intertidal zone to the edge of the continental shelf and includes the coastal waters having an average depth of 200 m. The water masses beyond the continental shelf and covering the deep abyssal plain constitute the **oceanic zone.** The depth of this zone is more than 3 Km.

Classification on the basis of light penetration

On the basis of distribution of light in the oceanic compartment, it is possible to distinguish two distinct zones namely: **(1)** Photic zone and **(2)** Aphotic zone. The **photic zone** is restricted up to that layer where light is available and usually ranges upto 200 m from the surface. The productivity is more in this zone and the producers include many species of large algae thriving in extensive beds, as well as huge population of phytoplanktons that serve as a nutrient base for the rest of the consumers. The **aphotic zone** is the deeper lightless zone of the ocean usually below 200 m.

1.3.3. Desert ecosystem

Precipitation is the limiting factor that determines whether the biomes of the world's land areas are desert, grassland or forest. A region will normally be a **desert,** if it contains very poor vegetation, widely spaced and if the average precipitation in the area is less than 25 centimeters (10 inches) per year. Many of the world's large deserts are found around latitudes 30° north and south of the equator, where the global climate system tends to produce descending masses of dry air. More than one-third of the earth's land surface is covered with deserts, where average precipitation is very low and evaporation rate is extremely high. The entire desert system of the earth may be divided into three broad groups **(1)** ***tropical deserts,*** such as the Sahara, where the temperature during the day time are hot year-round and nights are cold, **(2)** ***temperate (Mid-latitude) deserts,*** where daily temperatures are warm most of the year and **(3)** ***cold (High-latitude) deserts,*** such as the Gobi, where winters are cold and summers are hot.

Tropical deserts, making up about one-fifth of the world's desert area, consists primarily of barren dunes covered with rock or sand. Temperate and cold deserts contain widely scattered thorny bushes and shrubs (acacia, mesquite, sagebrush, greasewood etc.), succulent plants like cacti and small fast growing wild flowers that bloom in spring or after a rare, brief, drenching rain. The desert plants are characterised by long root systems for dragging water from a very low water table below the surface. The desert ecosystem sustains a variety of fauna like rodents, reptiles, ground squirrels, jackals, foxes, camels, bats, owls, and insects. The silica content of the desert soil is high and the concentration organic carbon is extremely low which are not favourable for the luxuriant floristic diversity of the ecosystem. Thus deserts represent fragile biomes with nutrient poor soils, slow growth rate of plants and extreme water scarcity.

1.3.4. Grassland ecosystem

Grasslands have flat or slightly sloping terrain that cannot support large stands of trees because of high evaporation rates, irregular precipitation, periodic drought and occasional fires (common in tropical and temperate grasslands). Grasslands are one of the most important terrestrial ecosystems, which occupy approximately 19% of the total world vegetation.

Tropical grassland or ***Savannahs*** are found in areas with high average temperatures, very dry seasons about half of the year and abundant rain in the rest of the year. Tropical grasslands may have 1,520 mm annual rainfall in wet dry season which alternates with a prolonged dry season. They are located in a wide belt on either side of the equator between the tropics. Some of these biomes, such as Africa's **Serengeti Plain**, consists of open plains having low or high grasses as the main producers, while others contain grasses along with varying numbers of widely spaced, small, mostly deciduous trees and shrubs, which shed their leaves during dry season to avoid excessive water loss. The primary consumers in African grassland ecosystem include gazelles, zebras, giraffes and antelopes. The secondary consumers like lions, leopards and cheetahs are vanishing rapidly from the area because of the threats like ranching, farming, hunting, poaching and other human activities.

In temperate grasslands, the average annual precipitation ranges between 255 mm to 750 mm. Also known as ***mid-latitude*** grasslands, these ecosystems are widely distributed in the large interior areas of continents, having moderate average temperature, more even rainfall than the tropical grasslands and wind blowing almost continuously throughout the year. Examples are the tall-grass and low-grass **prairies** of the mid-western and western United States and Canada, the ***pampus*** of South America, the ***veld*** of southern Africa and the ***steppes*** that stretch from Central Europe to Siberia.

Polar grasslands or high-latitude grasslands or Arctic ***tundra*** are concentrated below the Arctic region having perpetual ice and snow. The climate is characterized by bitter coldness, icy gale-like winds, fairly low average annual precipitation occurring primarily during a brief summer period and a long winter darkness. The wet Arctic tundra is covered with a thick spongy mat of plants like lichens, sedges, mosses, grass and woody shrubs that have usually low growth rates.

One important phenomenon related to extreme coldness in the polar grasslands is the presence of ***permafrost*** in which the water gets permanently frozen year-round in thick underground layers of soil. During the period of brief summer, the surface layer of soil thaws and the biome is turned into a sloggy landscape dotted with shallow lakes, marshes, bogs and ponds. The permafrost below prevents this water from soaking into the ground and the evaporation rate is reduced. This period witnesses an increment of biodiversity of the biome as the hordes of mosquitoes, deerflies, blackflies and other insects that thrive in the shallow waterbodies invite a large colonies of migratory birds, especially water-fowl, that fly in from the south.

Most of the tundra's permanent animal residents are small herbivores such as lemmings, hares, voles and rabbits, which burrow under the ground to escape the cold. Few species are present in large number which is an important feature of stressful environment.

The lows rate of decomposition by microbes, the shallow soil and the slow growth rate of plants make the arctic tundra perhaps the earth's most fragile biome.

The grasslands of the Indian -subcontinent have been classified by Whyte (1957) into eight types on the basis of floristic composition. These are: **(1)** ***Schima - Dichanthium*** type on black soils, **(2)** ***Dichantium- Cenchrus*** type on sandy loams, **(3)** ***Phragmites- Saccharum*** type on marshy localities, **(4)** ***Bothriochloa*** on paddy tracts and high rainfall belts, **(5)** ***Cymbopogon*** type on low hills, **(6)** ***Arundinella*** type on high mountains, **(7)** ***Deyeuxia- Arundinella*** type in mixed temperate climate and **(8)** ***Deschampsia-Deyeuxia*** type in mixed temperate alpine climate. The first four types are the characteristics of the plains and the last four types of the hills and mountains.

1.3.5. Forest ecosystem

Forest is an important ecosystem of the biosphere in respect of its standing crop biomass and its role in controlling the global biogeochemical cycles. A wide spectrum of vegetations representing stages in succession and adaptations to varying soil conditions may be noted in any forest region.

Tropical rain forests are found near the equator in areas with a warm climate having an annual mean temperature about 27°C, high humidity and considerable rainfall. Although such forests occupy only 7% of the earth's land surface, they contain 43% of all vegetation in the biosphere and constitute almost half of the earth's remaining forests. The climatic conditions do not fluctuate much in this ecosystem, and therefore neither water nor temperature are limiting factors as in other biomes. Because of this lack of major limiting factor, a mature tropical rain forest has a greater species diversity per unit area than any other biome.

Temperate (mid-latitude) deciduous forests occur in areas with moderate average temperatures that change significantly during four distinct seasons. They have long summers, not very severe winters and abundant precipitation throughout the year. They are dominated by few species of broad -leafed deciduous trees like oak, maple, poplar, hickory etc. The dominant herbivore of most deciduous forests in the eastern United States is the white tail deer. These forests invite a variety of migratory bird species like warblers, robins etc.

Cold (high-latitude) northern coniferous forest, also called boreal forests or taiga, are found in regions with a subarctic type of climate, where winters are long and dry with only light snowfall, temperatures range from cool to extremely cold, and summers are short with warm temperatures. These forests are widely distributed in northern part of America and Eurasia and are dominated by coniferous trees like spruce, fir, larch and pine. The primary consumers in these forests mainly include large herbivores like mule, deer, elk, caribou and moos. Small herbivores like porcupine, hares, squirrels are also very common. The secondary consumers or carnivores of these forests are grizzly bears, foxes, lynxes and wolves. A wide range of insects and worms are important biotic components of these forests whose diversity usually shows an increasing trend during the warm summer months.

1.4. ECOLOGICAL NICHE

Charles Elton (1927) in England was one of the first to use the term **niche.** Grinnell (1928) inclined on the term **microhabitat** to describe the niche concept. The habitat of an organism is the place where it lives and the term solely encompasses the physical space or the dwelling space of the organism. The ***ecological niche*** on the other hand, is a more informative concept that includes not only the physical space occupied by an organism, but also its functional role in the community (as for example its trophic position in environment gradients of temperature, moisture, pH, soil parameters like soil grain size, organic carbon and other conditions of existence). The three aspects of ecological niche can be conveniently designated as the ***spatial or habitat niche,*** the ***trophic niche*** and the ***multidimensional*** or ***hypervolume niche.*** Thus ecological niche of an organism signifies not only its dwelling place, but also its position in the food web which is an indicator of its efficiency in energy transformation and also how it is constrained by other species.

It has become axiomatic that *no two species can occupy the same niche coordinates indefinitely.* If even it happens, then their efficiency to adapt with the physico-chemical variables of the niche are different, which would eventually results in the suppression of one species (that which is not adapted in a better way) by the other. Niche segregation is an important step to reduce the competitive relationship between species. Within the same species, competition is often greatly reduced when different stages of the life cycle of an organism occupy different niches, for example the tadpole is an herbivore and the adult frog is a carnivore in the same pond or a caterpillars feeds directly on leaves, pupa ceases eating (minimum energy conversion) and the butterfly seeks for the honey of the flowers in the same plant. The same mangrove plant (*Avivcennia marina*) offers shelters to two different species of the gastropod *Cerithedia* in two different planes. *Cerithedia cingulata* are widely distributed at the muddy base, while *Cerithedia obtusa* at the vertical stem of the tree. This is a very common view while traversing the mangrove ecosystem of deltaic Sundarbans in the Indian sub-continent.

1.5. ECOLOGICAL SUCCESSION

In ecology, **succession** may be defined as the progression of organisms during the developmental history of an ecosystem.

All communities in this earth undergo a cycle of birth, growth and senescence like that in an individual organism. The difference is that communities do not die, rather one group of plants and animals replaces another. A community comes into existence when some barren habitat (like sandy beach, bare rock, volcanic island or mud flat) is first invaded by a group of ***pioneer*** species. These species modify the habitat so that other species can invade and persist. Finally the area is occupied by species that dominate the community and replace themselves, rather than being replaced by other species. This is a ***climax community***. This entire process of replacing groups of species, one by another, through habitat modification is called succession. The three important features of ecological succession are **(1)** It is a sequential phenomenon involving changes in community structure or species diversity with the passage of time **(2)** It is the result of the modification of the physical environment by the community **(3)** The process ends in a stabilized ecosystem in which the efficiency of energy conversion (into biomass) is maximum.

The process of ecological succession through formation of different types of vegetation in a definite order has been nicely demonstrated by Chaudhuri & Choudhury (1994) in mangrove ecosystem of the Indian Sundarbans. The usual pattern of succession in the deltaic region is as follows:

Porteracia coarctata

⇓

Agiceras corniculatum,
Avicennia alba, Avicennia officinalis,
Sonneratia apetala

⇓

Bruguiera sp.

⇓

Heritiera fomes, Rhizophora sp.
Xylocarpus molluscensis

⇓

Ceriops decandra, Exoecaria agallocha

Chapter 2:

ENERGY FLOW

2.1. ENERGY FLOW THROUGH FOREST ECOSYSTEM

Energy is defined as the capacity to do work. The behaviour of energy is associated with two important laws of thermodynamics. The **first law of thermodynamics** states that energy may be transformed from one type into another, but is never created or destroyed. The **second law of thermodynamics** expresses the idea that no energy conversion in the earth is 100% efficient. When the solar radiation strikes the earth, it tends to be degraded into heat energy. Only a very small portion of the light energy absorbed by green plants is transformed into potential or food energy. Most of the energy is wasted in the form of heat. All the rest of the biological world (consumer) obtains its potential energy from organic substances produced by plant photosynthesis or micro-organism chemosynthesis. An animal intakes the chemical energy of food and converts a large part into heat to enable a small part of the energy to be re-established as the chemical potential energy of new protoplasm. Thus the process of energy flow through an ecosystem is **unidirectional** and during each step of the transfer of energy from one organism to another, a large part of energy is degraded into heat.

The entire process of energy flow through an ideal forest ecosystem may be discussed under three broad headings:

Acquisition of energy by the producers

The Sun is the main source of energy in the forest ecosystem.

Extraterrestrial sunlight reaches the biosphere at a rate of 2 gcal per cm^2 per min, but is attenuated as it passes through different layers of atmosphere. According to Gates (1965) about 67 percent (1.34 gcal per cm^2 per min) may reach the earth's surface at noon on a clear summer day. The magnitude of radiation penetrating the atmosphere is a function of several factors like atmospheric gases, dust, wavelength of the radiation, etc. Shortwave ultraviolet radiation below 0.3 μ is abruptly terminated by the ozone layer in the outer atmosphere (about 18 miles or 25 km altitude). Solar radiation is further attenuated and the spectral distribution of its energy is greatly altered as it passes through the cloud cover, water and vegetation. Radiant energy reaching the surface of the earth on a clear day consists of 10 percent ultraviolet, 45 percent visible and 45 percent infrared (Reifsnyder and Lull, 1965). Visible radiation is least attenuated as it passes through dense cloud cover and water, which means that photosynthesis can occur even on cloudy day at some depth in clear water. This concept is very important for understanding the acquisition of energy by aquatic plants which are present in the waterbodies of moist deciduous forests in the tropics. The daily input of sunlight to the autotrophic layer of an ecosystem varies mostly between 100 and 800 gcal per cm^2 averaging about

300-400 gcal per cm^2 for an area in the temperate zone such as the United States (Reifsynder and Lull, 1965). It has been observed by Gates (1965) that vegetation of the earth strongly absorbs the blue and red visible wave lengths, less strongly the green wave length and very weekly the near infrared. The blue and red light (0.4-0.5 μ and 0.6-0.7 μ bands respectively) are especially absorbed by chlorophyll and the far infrared heat energy is absorbed by the water in the leaves and the water vapour that surrounds them. Green plants thus efficiently absorb the blue and red light that are most useful in photosynthesis.

Fixing of energy by the producers

The flow of energy through the forest community starts with the fixation of sunlight by plants, a process that in itself demands the expenditure of energy. Initially plants rely on the food stored in the seed for energy until their own production machinery is working. Once mobilized, the green plant begins to accumulate energy. Energy accumulated by plants is called production, or more specifically, **primary production,** because it is the first and basic form of energy storage in an ecosystem. The rate at which energy is stored by photosynthesis activity is known as **primary productivity**. All of the sun's energy that is assimilated (i.e., total photosynthesis) is **gross primary production**. Like other organisms, plants require energy for reproduction and maintenance. The energy required for these needs is provided by a reverse of the photosynthetic process, respiration. The energy remaining after respiration and stored as organic matter is **net primary production**, or plant growth. Thus net production for a single plant or a plant community can be described by the formula

Net primary production =
Gross production – Autotrophic Respiration

The basic unit of energy used in ecological work is the calorie; production is usually expressed in kilocalories per square meter per year (Kcal/m^2/year). However, production may also be expressed as dry organic matter in grams per square meter per year (g/m^2/year).

Net primary production accumulates over a period of time as plant biomass. Part of this accumulation is turned over seasonally through decomposition. Part is retained over a longer period as living material. The amount of this accumulated living organic matter found on a given area at a given time is the **standing crop biomass**. Biomass is usually expressed as grams dry weight of organic matter per square meter (g/m^2), kilocalories per square meter (kcal/m^2), or some other appropriate measure per unit area.

Different types of forests ecosystems have characteristically different productivities depending on the quality and quantity of producers. The net primary productivity (NPP) of tropical rain forest ranges between 1000-3500 gm.cm^{-2}year^{-1}. The mean values of NPP for tropical seasonal forest, temperate evergreen forest, temperate deciduous forest and boreal forest are 1600, 1300, 1200 and 800 gm.cm^{-2}year^{-1} respectively.

In the Indian sub-continent, the forest primary productivity accounts to a considerable value owing to the presence of diverse type of plant species in diverse category of forest.

In **tropical rain forest** (which is influenced by the south-west monsoon) of the West coast extending from the

North Kanara to the South and in the eastern sub-Himalayan tract in Assam, the producers include many families among which Dipterocarpaceae, Guttiferae, Lauraceae, Magnoliaceae, Euphorbiaceae and Palmae are dominant. Climbers and epiphytes are plentiful among the producers of the ecosystem. The producers of the **moist deciduous forests** in India mainly consists of bamboo, teak (*Tectona grandis*), sal (*Shorea robusta*) and sandalwood (*Santalum album*) of Karnataka. In India such forests are concentrated along the western side of the Deccan plateau i.e., Bombay, North-east Andhra, Madhya Pradesh, Gangetic plain and some areas of the sub-Himalayan tract, extending from the Punjab in the West to the Assam valley in the East. **Dry or temperate deciduous forests,** in India, occur in the northern Punjab where the rainfall usually ranges between 50 cm to 75 cm. The important producers in this ecosystem are *Acacia arabica, Acacia modesta, prosopis spicigera, Albizzia lebbek* etc. The dry deciduous forests in Tamilnadu are dominated by *Soymida febrifuga, Terminalia* spp., *Tectona grandis* etc.

The net primary production of any forest ecosystem is not only accounted by the vegetative biomass above the ground, but the root systems below the soil surface also have contribution to the productivity.

Transference of energy through trophic level

The transfer of food energy from the source in the plants through some tiers of organisms (each tier is referred to as the trophic level) constitute the food chain of an ecosystem. In forest ecosystem many food chains, being intricately related to each other, form the food webs. All the consumers of any type of forest ecosystem (starting from squirrel to lion) depend on the value of net primary productivity for their supply of energy. However, ecosystems differ in the *efficiency* with which consumers at different trophic levels are able to acquire energy from the level below them and to incorporate that energy into *biomass*, the weight of the living organisms. Many kinds of ecological efficiencies have been given diverse names (Kozlovsky, 1968).

Lindeman's efficiency, also called *ecological efficiency* is the ratio of energy acquired by one trophic level (N+1) to that acquired by the previous (N). *Assimilation efficiency* is the fraction of light striking plants that is actually used in photosynthesis, or it is the fraction of energy consumed by animals which is absorbed form the food. *Growth efficiency* pertains to the splitting of the assimilated energy into the portion degraded through the metabolic activities of organisms or lost through excretion and the portion incorporated into new tissues. *Exploitation efficiency* pertains to the fractions of those tissues that are consumed by the next trophic level. Lindeman's efficiency is the product of assimilation efficiency, growth efficiency and exploitation efficiency.

Assimilation efficiency is a unique function of diet spectrum of an organism. For example, a millipede feeding on decaying wood of a forest tree may have an assimilation efficiency of 15% (O'Neill, 1968). An elephant depending on vegetative parts of the forest plants may have about 30% (Pelrides and Swank, 1966). A seed-eating Kangaroo rat may have about 80% (Chew and Chew, 1970). In general carnivores are more efficient than herbivores due to the presence of much more indigestible material (like cellulose) in the plant tissues. Growth efficiency at the producer level can be quite high, in contrast to assimilation efficiency. Plants do not spend a lot of energy in locomotion and thus are able to incorporate a high percentage of material in their tissues. An efficiency of 30% has been estimated for a tropical rain forest (Golley and Misra, 1972). Efficiencies above 60-70% are common, especially outside of the tropics.

Generally ectotherms have higher growth efficiencies than endotherms as the former type do not divert energy towards maintaining their body temperatures. Insects tend to use about 33% of their assimilated energy in tissue building and reproduction, birds and mammals less than 1-5%. Growth efficiencies tend to be higher in young than in old animal (Phillipson, 1966).

The generalised pattern of energy flow often becomes complicated when some forest animals like leopards are considered for case study purpose. This is because their diet chart includes a wide spectrum of animals and hence the energy from different consumer compartments converge to a single compartment resulting in an amplification of energy in the animal body. The leopard scat analysed from North Bengal area at the base of the Himalayan range during 1996-97 revealed the presence of 18 different species (namely *Suncus maximus, Macaca mulatta, Macaca assamensis, Presbytis entellus, Muntiacus muntjak, Axis porcinus, Tragulus meminna,* Domestic goat, *Sus scrofa, Paradaxurus hermaphroditus, Bandicoota indica, Ratus ratus, Calloscirus pygarythrus, Cannomys badius, Petaurista petaurista, Ratufa bicolor* and *Gallus gallus*). This animal thus presents another model of energy flow in which energy of the consumer does not decrease to that extent (in comparison to the animal that feed on single species) as the energy paths from a wide range of prey species find their way into the predator. However during all cases of energy transference through trophic levels in forest ecosystem Lindeman's 10% rule is conveniently applied.

Lindeman's efficiencies are often said to be about 10% at levels above the producer level. This 10% rule of thumb is convenient for peeping into various ecological problems. Wc can visualize each successive trophic level as having roughly one-tenth of the energy of the preceding level available to it. On this basis it can be stated that with the same primary production about 10 times more biomass of deer who eat grass could be supported than biomass of tiger who depend on deer in thc forest ecosystem. This progressive decrease of available energy as it passes along food chains, following the dictates of the second law of thermodynamics, results in a pyramidal structure of ecosystems. Figure 1 highlights the pyramid of biomass for a tropical forest.

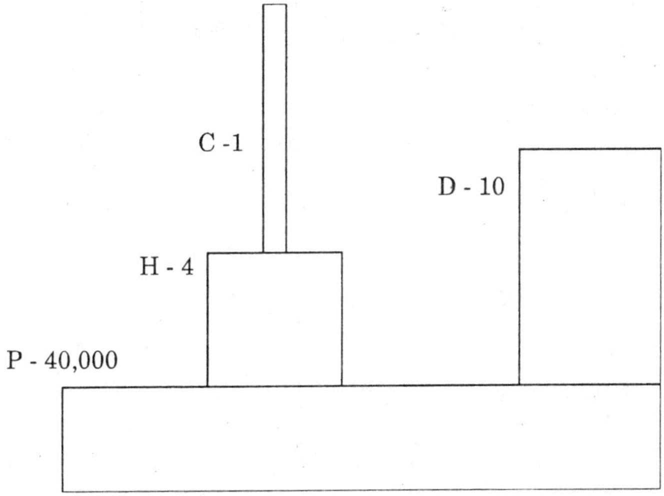

Fig. 1: Pyramid of biomass of Panama tropical forest (After E. Odum, 1971). P = producer; H = herbivores; C = carnivores and D = decomposers. Numbers represent Individuals per $1000m^2$.

2.2. ENERGY FLOW THROUGH AQUATIC ECOSYSTEMS

The main factors affecting the biological diversity of aquatic ecosystems are salinity, transparency of the aquatic phase (or the depth up to which sunlight can penetrate), dissolved oxygen and water temperature. **Salinity,** the concentrations of dissolved salts (especially sodium chloride) in aquatic phase is a limiting factor because different species of aquatic plants and animals have different salinity tolerances. Salinity levels are used to divide aquatic ecosystems into two major classes **(1) freshwater ecosystems** consisting of inland bodies of standing water (like lakes, reservoirs, ponds etc.) and flowing water (like streams and rivers), and **(2) marine or salt water ecosystems,** such as oceans, estuaries, coastal wetlands and coral reefs.

The basic mechanism of energy flow starting from the producer level through different trophic tiers is same in all types of ecosystem, but due to some factors like depth, salinity, upwelling, nutrient load etc., the quantum and nature of primary production greatly differs among aquatic ecosystems.

Freshwater ecosystems

Ponds are small, shallow usually human created impoundments of freshwater where photosynthesis can take place at all depths during daylight unless the surface water becomes clogged with masses of algae during warm weather. In this ecosystem, the energy flow is triggered by various algae like *Chlamydomonas, Volvox, Spirogyra* etc. The primary consumers include variety of zooplankton and herbivorous fishes and they transfer the energy further to higher trophic levels consisting of large fishes, crabs, birds (like kingfisher) etc.

Streams are relatively small, flowing waterbodies that empty into rivers. These waterbodies receive a considerable contribution from the adjacent lands in the form of run-off and nutrient depletion is not witnessed through out the year. Excessive addition of nutrients and organic materials to streams and other waterbodies can cause undesirable effects on secondary production, but up to certain level the addition of small amount of nutrients to aquatic ecosystems can stimulate desirable secondary production. Waren *et al.* (1964) showed this by the experimental addition of sucrose to Berry Creek in Oregon. It was found that trout productions were very high in the sucrose-enriched section in the creek rather than the control section. The abundance of benthic insects in the sucrose enriched sections also increased due to the increased availability of the bacterial food in the form of a slime bacterium, *Sphaerotilus natans.* He pointed out that not all energy pathways through this stream community lead to trout production. Trout depend only on a few selective species of benthic invertebrates whose abundance increased due to addition of small amount of sucrose. However, if sucrose addition were increased greatly, the population of benthic invertebrates would be under severe stress as the colony of slime bacteria would smother the bottom. In such case, the production of trout would not exhibit an increasing trend. The principle of employing critical or optimum level of nutrients to initiate the process of energy flow is also practiced since a long period for farming and pisciculture in and around the East Calcutta Wetlands. The sewage discharged from the highly urbanised and industrialised city of Calcutta, being packed with nutrients like nitrate and phosphate, is diverted into the waterbodies of East Calcutta Wetlands where these nutrients initiate the growth of phytoplanktons. These planktonic communities, being the main players behind the primary production of the aquatic ecosystems, play a great role in the production of fishery. The important fishes constituting the consumer level of these wetlands mainly include *Channa punctatus, Channa striatus, Channa gachua, Labeo rohita, Catla catla, Cirrihinus mrigala, Hypopthalmicthys molitrix, Cyprinus carpio, Tilapia mosambica, Liza parsia, Liza tade, Lates calcarifer, Oreochromis nilotica, Clarias batrachus, Heteropneustes fossilis, Mystus gulioi, Puntius japonica* and *Anabas testudineus*. A survey done to investigate the interrelationship between NPP and fish yield at 15 waterbodies in the East Calcutta Wetlands during 1995-96 revealed that the value of NPP of the waterbodies receiving sewage ranges between 21.7 mg C m^{-3} h^{-1} to 50.1 mg C m^{-3} h^{-1} and that of fish yield between 1.5 tons/Ha to 7.0 tons/Ha. A unique positive correlation between NPP and secondary production ($p<0.01$) was also revealed from the study.

Marine ecosystem

The mechanism of energy flow in the marine ecosystem is bit complicated as life in the oceans is unevenly distributed. The **neritic zone,** representing 10% of the total ocean area, contains 90% of all ocean plant and animal life and is the site of the commercial marine fisheries. The primary productivity due to solar energy in the marine ecosystem is limited in the photic zone (approximately upto 200 m deep from the surface) as the sun rays fail to penetrate beyond this zone. This is one of the main reason why a huge open sea has a relatively low average net primary productivity of plant life.

The nature of energy flow in marine ecosystem differs in different zones of compartments of ocean. The neritic or coastal zone is relatively warm, nutrient rich shallow water zone that extends from the high tide mark on land to the edge of shelf-like extension of continental landmasses known as the continental shelf. Estuaries are distributed along the coastlines, where there is a continuous overlapping of fresh and saline water due to tidal actions. All these zones are the sites of high primary production owing to the presence of a variety of producers like **(1)** sea grasses (*Zostera* sp.) **(2)** salt marsh grass (*Porteracea coarctata*) **(3)** mangroves (*Avicennia* spp., *Rhizophora* sp., *Heritiera fomes* etc.) **(4)** sea weeds (*Ulva* sp., *Enteromorpha* sp. etc.) and **(5)** Phytoplankton (*Coscinodiscus* sp., *Bidulphia* sp. etc.). The estimated average net primary productivity by producers in estuaries is about 9,200 Kcal/m^2/yr, but when the average net primary productivity is compared among the major types of ecosystem of the world, it becomes extremely low (about 18 billion Kcal/yr.), whereas for open ocean the value becomes 200 billion Kcal/yr. This data often becomes misleading as the productivity in open ocean is very much less in comparison to estuaries and coastal wetlands. However, as the oceans cover so much of the globe in terms of area, they head the list of NPP among the major types of ecosystem in the world although the productivity value per unit area per unit time is extremely poor. The energy from these producers pass to the zooplankton, who mostly graze over the phytoplankton for building their own biomass. The various fin fishes, crabs, oysters, squids and birds like gulls are the top level consumers of the marine ecosystem. The coastal region exhibits a unique reservoir of nutrients not only because of their derivation from the adjacent landmass through run-off, but also due to some processes like upwelling. Along some steep coastal areas, such as those of Peru, winds blow the surface water away from the shore, allowing cold, nutrient-rich bottom waters or upwellings, to rise to the surface and support large population of plankton. In areas of upwelling, food chains are usually short. This is primarily because of the large size of the phytoplankton, many of which form colonies several millimeters in diameter. These large colonies are directly consumed by herbivorous fishes like sardines, anchovies etc. making the food chains of the area relatively shorter. Thus the process of energy flow in neritic or coastal zone is governed greatly by wind driven currents.

The sharp increase in the depth of the water at the edge of the continental shelf marks the separation of the neritic zone from the **open ocean**. This compartment snatches about 90% of the total surface area of the ocean, but sustains only about 10% of its plant and animal life. This is mainly because light is available only at its surface layer and most nutrients are found on the bottom far below.

The open sea can be broadly divided into three distinct vertical zones on the basis of magnitude of NPP. The surface layer through which enough sunlight can penetrate for photosynthesis is called **euphotic zone.** Phytoplanktons are the main primary producers of this zone. The phytoplanktons comprise a variety of organisms namely diatoms, dinoflagellates, blue green algae, silicoflagellates, coccolithophores etc. which have a size range of 0.001 to 0.2 mm and contributes to about 95% of the primary production in the open sea eutrophic compartment. The diatoms commonly known as the *pasture of the sea* or *meadows of the* *sea* are unicellular autotrophic organisms and form about 90% of the phytoplankton biomass. The zooplankton community of the euphotic zone being sustained by the phytoplankton biomass, support commercially important fishes like herrings, sardines, anchovies and other small fish that feed at the surface and larger predators such as tuna, mackerel, sword fish etc.

The quantum of input energy, however, decreases in the bathyal zone, which is a colder, darker layer below the euphotic zone. As the light penetration is very scanty in this layer, therefore the photosynthetic activity is greatly hindered. The pitch dark, colder abyssal zone is witnessed as one goes deeper towards the bottom. In the absence of photosynthesis, most of the organisms in this layer survive by feeding on dead plants and animals and their waste products, which sink down from the surface waters, and also by making negatively geotropic migration to surface layer for the purpose of procuring food materials. Thus the transference of energy in the aphotic depth of the ocean is not a smooth process and due to cut down of the initial input energy, the NPP and species diversity in this zone is extremely low.

A comparative picture of the estimates of primary production and fish production in three different zones of the oceans (Table 2.1) clearly reveals the low productivity and low efficiency of energy transfer in the open ocean in comparison to the coastal zone.

In 1977, a unique aquatic ecosystem was discovered deep in the abyssal zone around vents in the ocean floor of the Pacific Ocean. John Corliss and his co-workers explored the Galapagos Rift, a volcanic active boundary between the two

Table 2.1: Estimates of primary production and fish production in three zones of the ocean (After Ryther, 1969).

Zone	Percentage of ocean (gC/m^2/yr)	Mean primary productivity	Total primary production (billions of tons C/yr)	Number of trophic levels	Efficiency of energy transfer	Fish production (tons wet weight)
Open ocean	90	50	16.3	5	10	0.16×10^7
Coastal zone	9.9	100	3.6	3	15	12.0×10^7
Upwelling areas	0.1	300	0.1	1 or 2	20	12.0×10^7
Total			20.0			24.16×10^7

earth's crustal plates and observed that in that zone, the near freezing seawater seeps into the fissures becomes heated and is spewed out through vents at temperature exceeding 700°F. This hydrothermal effluent deposits zinc, iron and copper sulphides as well as calcium and magnesium sulphates, all leached from rocks as pressure forces the heated waters upward. In contrast to most of the deep oceanic compartment, which is often called *biological desert,* these nutrient rich *warm water oases* support diverse marine communities whose foundation lie on the unique type of energy source. In this zone chemosynthetic bacteria and other microbes use the inorganic deposits as energy sources. They are the primary producers in the trophic level whose higher tiers are occupied by tube worms, clams, crabs and fishes.

2.3. CYCLING OF INORGANIC NUTRIENTS

All inorganic substances essential for the life on earth tend to move in cycles from the physical environment, to organisms, then back to the environment. These circular movements of the essential inorganic substances are referred to as **biogeochemical cycles.** Living organisms require six elements in relatively large quantities: carbon, hydrogen, oxygen, nitrogen, phosphorus and sulphur. Elements needed in smaller amounts include sodium, potassium, manganese, calcium, iron, magnesium, chlorine, iodine, cobalt and boron. These elements are present in rocks, usually as salts. They are released, by erosion and weathering, into soil, rivers, lakes and the oceans. Some elements, such as nitrogen and oxygen, are also present in the atmosphere. The movements of nutrient elements through the biosphere, or though any particular ecosystem, by the physical processes of erosion, deposition, evaporation and precipitation and by the biological processes of uptake and release constitute the biogeochemical cycles. They are called cycles because nutrient elements, unlike solar energy, may be used over and over again by living systems.

Nutrients are sometimes recycled rapidly through living systems, as in grasslands, where the above ground vegetation dies away every year and its nutrients are made available again. In some cases, nutrients are locked for many years apart from the activities of the biological world. For example, remains of marine organisms may sink to the bottom and get incorporated into sedimentary rocks that are uplifted and exposed to erosion and nutrient release only after millions of years. Every nutrient element has a somewhat different fate, depending on its physical and chemical properties and on its role in living organisms. In the present text, the concept of nutrient cycling is discussed below with only a few simplified examples.

Carbon Cycle

In terrestrial ecosystem, there is a short term cycling of carbon between living organisms and the atmosphere, as a result of photosynthesis and respiration. Most of the carbon fixed as organic matter by the producers is rapidly broken down and released back into the atmosphere as CO_2. In aquatic ecosystems, the corresponding exchange is between living organisms and CO_2 or HCO_3^- (bicarbonates) dissolved in the surrounding water. In some ecosystems, however, carbon may accumulate either as undecomposed organic matter, as in peaty layers of bogs and moorlands, or as insoluble carbonates, which accumulate in bottom sediments in aquatic systems. Such carbon has entered a cycle of longer duration, and may take a long period to be released. Deposits of coal, oil, natural gas represent part of the net production of a bygone era. These fossil fuels are organic compounds that were buried before they could be decomposed, and were subsequently transformed by time and geological processes. As they are burned, their carbon is released back into the atmosphere as CO_2. Sedimentary carbon eventually turns into sedimentary rocks such as limestone and dolomite. After millions of years, these rocks may be lifted above sea level and exposed to erosion, releasing their carbon as carbonates (CO_3^{2-}) and bicarbonates into streams and rivers.

The carbon balance of the biosphere as a whole is moderated by exchange of CO_2 between the atmosphere and the oceans. The CO_2 content of the atmosphere is about 0.03%, but is steadily increasing due to excessive burning of fossil fuels and clearing down of vegetations, both of which are very keenly related to industrialization and urbanization. The rate of CO_2 increase in the atmosphere, however, is less than half of what would be expected from these human activities. The remainder is almost certainly absorbed by the oceans, which thus act as a global **buffer** for CO_2, moderating its concentration in the atmosphere and accounting for the basic stability of the global carbon budget.

Nitrogen Cycle

The movement of nitrogen through an ecosystem differs from that of carbon in that most of the enormous pool of molecular nitrogen gas is available only to those bacteria and cyanobacteria that can fix gaseous nitrogen by reducing it to ammonia. Plants acquire this vital nutrient either directly from nitrogen-fixing micro-organisms or indirectly from the limited pool of nitrogen found as ammonium or nitrate ions in soil water, rivers, lakes and oceans. Other inputs to this pool include nitrate formed from atmospheric nitrogen by lightning, ammonium or nitrate from the decay of organic matter, and nitrate from the erosion of nitrate-rich rocks. Certain bacteria and fungi use the proteins and amino acids in dead organic for their own dietary nitrogen, releasing the excess as ammonia or ammonium ions, some of which is again taken up by plants.

Nitrifying bacteria, also in the soil, oxidize some of the ammonia to nitrite. Finally, there are other bacteria that oxidize nitrite to nitrate, another form of nitrogen that can be absorbed by plants.

Nitrogen fixation requires a great deal of energy, which nitrogen-fixing micro-organisms must obtain from organic matter. Some decompose organic matter in the soil, some live as symbionts in the roots of higher plants and some are photosynthetic. Thus nitrogen fixation occurs at a slow rate relative to the rate of uptake of fixed nitrogen by plants. Consequently, despite the fact that nitrogen makes up 78% of the atmosphere, it is often the limiting nutrient in plant growth. Due to this reason nitrogen is most commonly applied in crop fertilizers.

Although many ecosystems, such as tropical rain forests, are remarkably efficient at retaining their nitrogen, almost all terrestrial ecosystems continually lose some nitrogen as nitrate and organic matter dissolved in ground water and streams. In the final phase of the cycle, nitrogen is returned back to the atmospheric "reservoir" by the denitrifying bacteria living in the anaerobic muds of fertile lakes, bogs, estuaries and on parts of the sea floor. The process of returning is not a sudden process, rather it takes place in a step-wise manner as shown below:

Protein (in flora and fauna)

⇓

Amino acids

⇓

Ammonia

⇓

Nitrite

⇓

Nitrate

⇓

Nitrogen

Phosphorus Cycle

Phosphorus is a major constituent of ATP, DNA, RNA and biological membranes and many animals need large quantities of this element for shells, bones, and teeth. Since phosphorus almost never occurs as a gas, its cycle, unlike the atmospheric cycles of nitrogen and carbon, is a sedimentary cycle. When rocks are eroded by weather, minute amounts of phosphorus dissolve, usually as phosphate, and become available to plants. Unlike nitrogen, which must be released from organic compounds before plants can take it up again, much of the phosphorus excreted by animals is in the form of phosphate, which is immediately reusable by plants. Thus cycling of phosphorus in terrestrial ecosystems is usually very efficient, although small amounts are continually lost down to aquatic phase like oceans. Since some phosphate salts are not very soluble in water, the oceans continually deposit phosphate minerals in their sediments. The main reason for the infertility of the open oceans is their low phosphorus content.

At the level of the biosphere, the phosphorus cycle is, in the short run, a one way flow from rocks to land ecosystem to the oceans from where it is finally transferred to the sediment compartment. Apart from trivial amounts deposited on land in the droppings of oceanic birds (Guano), the only natural way for phosphorus to return to land is via immensely slow geological processes in which sea floor sediments may again become terrestrial rocks. However, terrestrial ecosystems are able to retain much of their phosphorus for short term cycling, since both organic humus and inorganic soil particles absorb phosphate ions, providing a local reservoir of this element in the soil and helping to promote a steady supply of it into and out of the living organisms of the ecosystem.

2.4. CYCLING OF ORGANIC NUTRIENTS

Organic nutrients usually encompass vitamins and other growth promoting compounds that are necessary for the survival and normal growth of the organisms. Heterotrophs and many autotrophs (like algae) require vitamins and other organic nutrients which they must obtain from their environment.

It has become evident that organic nutrients play an.

important role in community metabolism and they can be also limiting factor.

In marine ecosystem, the main producers of vitamin B_{12} (cobalamins) are the micro-organisms mostly bacteria. Autotrophic algae are also important producers of vitamins in the marine environment, either as direct producer or, after their death, as food for vitamin producing micro-organisms. The consumers of vitamins in the marine and estuarine ecosystem include the filter feeders having extensive and highly permeable gills. These gills absorb vitamins directly as solutes. It has been observed that changes in size occurs when there is a shifting of the nutrient requirement from inorganic to organic (Odum, 1971). In a mineral and nutrient rich environment, small size is of selective advantage, especially to autotrophs because of the greater surface-to-volume ratio. As the ecosystem develops, however, inorganic nutrients tend to become more and more tied up in the biomass (that is, to become intrabiotic), so that the selective advantage shifts to larger organisms (either larger individuals of the same species or larger species or both), which have greater storage capacities and more complex life histories. They are thus adapted to exploit seasonal or periodic releases of nutrients or other resources. With this initiation of the releasing process, the load of organic nutrient in the environment increases.

2.5. CONCEPT OF HOMEOSTASIS

Homeostasis refers to the tendency of biological systems to resist change and to remain in a state of equilibrium. It has therefore three important components: sensor, effector and a control unit. The heat regulation in *Homo sapiens* is an example of homeostasis in biological sphere. In human being heat is felt by the receptor of the skin. This information is ultimately transmitted to the brain which involuntarily increases blood flow through the skin, thereby cooling the body. As the skin cools, the information is immediately fed back to the receptor system which shuts of blood flow to the skin and stops sweating.

The concept of homeostasis can be suitably explained through the familiar household heat control system, where the thermostat is the sensor, the furnace effector and the room temperature the parameter to be controlled. Control depends on feedback, which occurs when output or part of it feeds back as input. When this feedback input is positive, the variable or the parameter grows or accelerates. However, in order to achieve the balanced state as for example, to prevent the room from getting overheated, a sort of negative feedback is also necessary, otherwise the system would collapse.

In human beings few important examples of homeostasis are listed below:

- Blood glucose concentration remains at about 0.1 percent.
- The pH of the blood is always near 7.4.
- Blood pressure in the brachial artery averages near 120/80.
- Blood temperature averages around 37°C (98.6°F).

Chapter 3:

POPULATION ECOLOGY

3.1. PROPERTIES OF POPULATION

Considered ecologically, a population is a group of organisms of the same species occupying a particular space. But more than this, a population is one unit through which energy flows and nutrients are cycled. It is a self regulating system that helps to maintain stability within the ecosystem. A population, as with any other level of organization, has a number of important group properties, not shared by adjacent levels (the organism on the one hand and the community on the other; Fig. 2).

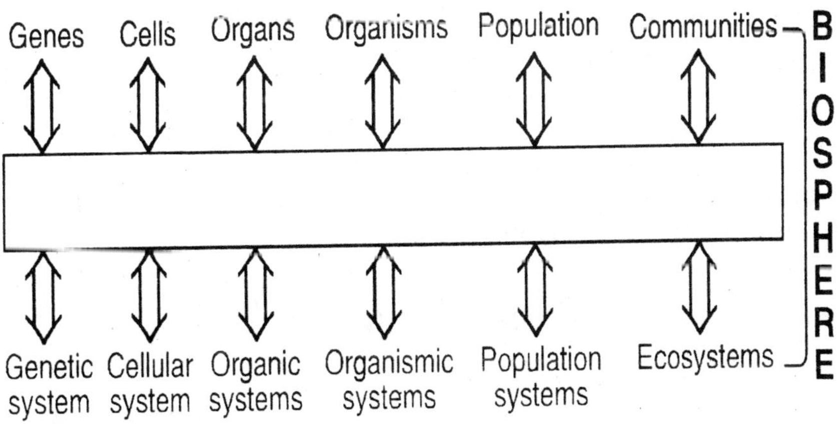

Fig. 2

The important population characteristics or group attributes include density, birth rate, death rate, population growth rate, dispersal, age structure and dispersion.

Density

It is the population size in relation to a unit space. Density are of two types: crude density and ecological density. **Crude density** is the total number of individuals or biomass per unit of total area or volume. It does not concentrate on the fact that whether the entire unit is a suitable habitat or not. However populations do not occupy all the space within a unit, because the entire space may not be a suitable habitat. A sample of soil, for example, may contain 2 million arthropods/m^2, but these arthropods do not inhabit the entire substrate. They live only in the pore spaces of the soil. Thus **ecological density** is the number of individual or biomass per unit area or volume which is actually inhabited by the individuals.

Birth rate

Birth rate or natality may be defined as the rate at which new individuals are added to the population by reproduction. It may be expressed as crude birth rate or specific birth rate.

Crude birth rate is expressed in terms of population size, e.g., 50 births/1000 population/year. Mathematically it can be expressed as follows:

$$\text{Crude birth rate} = \frac{\Delta Nn}{\Delta t}$$

Where ΔNn = new individual, and Δt = change in time.

Specific birth rate is expressed relative to specific criterion, such as age or condition. The usual form is the number of individuals produced per unit time by a female of age "X", since the reproductive capacity varies with the age. Mathematically, it can be expressed as follows:

$$\text{Specific birth rate} = \frac{\Delta Nn}{N\Delta t}$$

Death rate

It is the rate at which the individuals are lost by death from a given set of population, and, has therefore great influence on the population size. It is often expressed as the number of deaths per 1000 individuals of a population per year. Lowest death rate for a given species in most favourable conditions is called **potential mortality**, while the actual death rate being observed in existing conditions is called **realized mortality.**

The percentage ratio of natality over mortality is called vital index which can be mathematically expressed as follows:

$$\text{Vital index} = \frac{Natality}{Mortality} \times 100$$

Population growth rate

Population size of any species is a function of natality, mortality and dispersal. Basically there are two contrasting ways by which population grows. In the graphical language, the two ways of growth are called as sigmoid growth form (Fig. 3) and J-shaped growth form (Fig. 4).

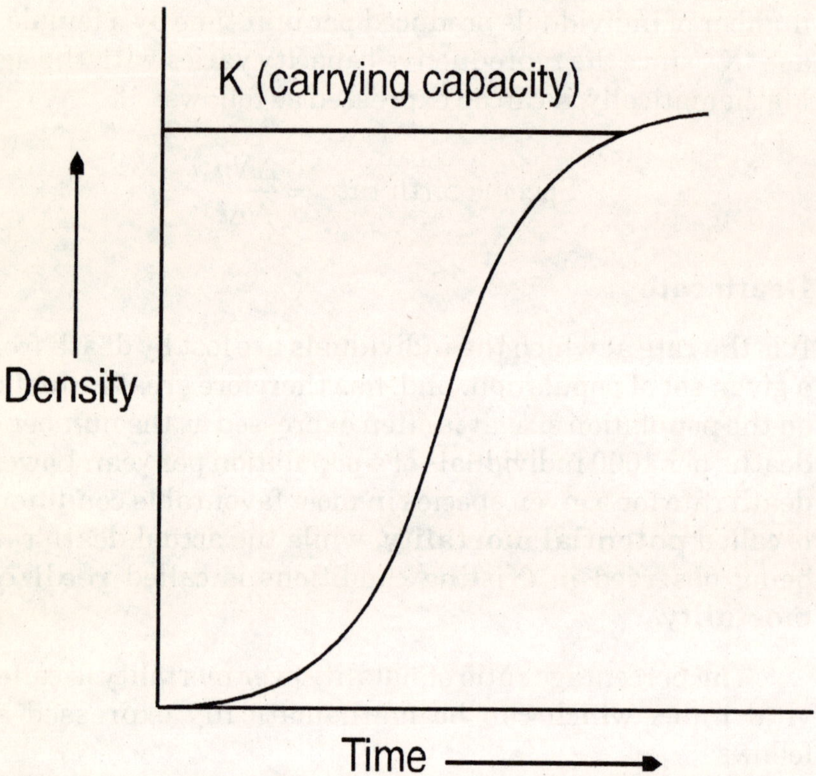

Fig. 3

The **sigmoid** or the "S" shaped growth curve consists of three parts namely: the early phase of little or no growth, the middle phase of rapid or geometric growth and the last phase of zero growth. The growth rate decreases towards zero as the population approaches the carrying capacity **K**, which is a measure of the feeding capacity of an environment or ecosystem for a population of species under a given set of conditions. However, in practical cases density often overshoots the carrying capacity, **K,** resulting in oscillations.

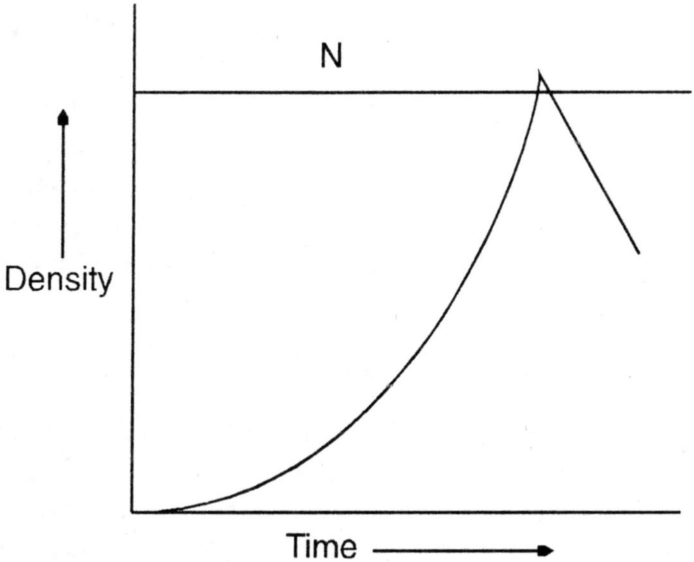

Fig. 4

The main feature of **J shaped** growth is the exponential increase of individuals of a population, e.g., 2, 4, 8, 16, 32 .... and so on, until the population runs out of some resources or encounters some other limitations (denoted by "N" in Fig. 4). This form may be represented by the model:

$$\frac{dN}{dt} = r_m N$$

Where, $\frac{dN}{dt}$ = rate of population growth, i.e., change in number per unit time, r_m = capacity for increase in a particular environmental conditions and can be assumed as instantaneous coefficient of population growth, and N = population size (number).

Dispersal

It is the rate at which the individuals immigrate into the population or emigrate out of the population. This property greatly affects the size of the population as a whole. Infact population dispersal is the movement of individuals or their disseminules or propagules (seeds, spores, larvae etc.) into or out of the population. In addition to immigration and emigration, the periodic departure and return of individuals (migration) is also a form of population dispersal.

Age structure

The age structure of a particular population refers to the population of individuals of various ages. Age is not only expressed in days, months or years, but it can also be expressed in the categories like pre-reproductive, reproductive and post-reproductive. The size of the population of any species is a direct function of the reproductive age group, since this age group adds up to the population through the process of reproduction. Basically a population can exhibit any one of the following three trends:

1. **Increase**: In this case there is a high production of the youngs.

2. **Decrease**: In this case there is a low production of youngs, but a large reproductive age group.

3. **Stable**: In this case the ratio between the youngs and the adults is approximately 2:1.

The age structures of the three possibilities are represented in Fig. 5.

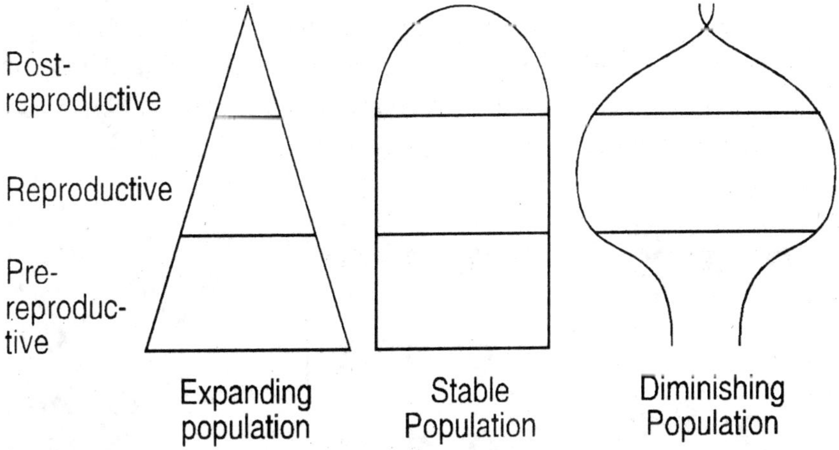

Fig. 5

Dispersion

It is a way by which the individuals are distributed in the space. There are three main types of distribution like random, uniform and clumped (Fig. 6).

Random

As the name indicates this type of distribution is a haphazard and irregular. In this type of distribution, the probability of finding an individual at any point in the area is same for all the points.

Uniform

In this type of distribution, the individuals are arranged in a regular manner e.g., distribution of tea plants in the tea gardens or the distribution of corn in a corn field.

Clumped

It is the most frequent type of distribution indicating the existence of individuals of a population in aggregations or groups e.g., a flock of birds, a herd of cattles, a colony of oysters or barnacles etc.

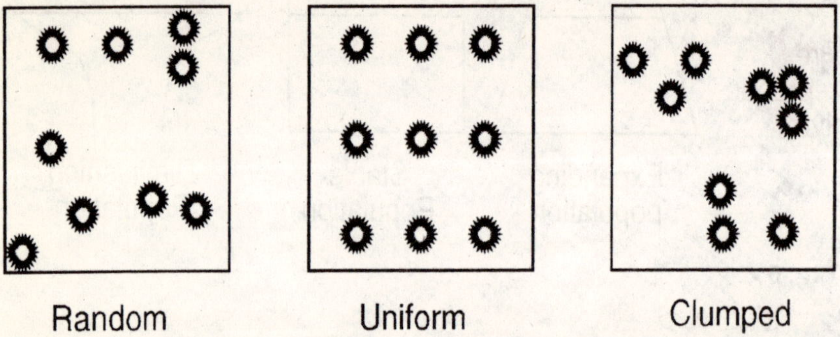

Fig. 6

3.2. POPULATION GROWTH FORM

All populations have the potential for rapid and enormous growth in numbers like the formation of lighter fragments from uranium as evidenced in chain reaction. However, question often peeps in mind that in what form will this increase occur and how it can be represented mathematically? For the purpose of comparative study, we may focus on two basic patterns of population growth namely the **J-shaped growth** form and the **S-shaped or sigmoid growth form**. These contrasting types may, however, suffer diverse modifications depending on the nature of the species and the environmental parameters. The two basic forms of population growth are discussed separately:

J-shaped growth form

In this form, the population density in an environment increases rapidly in exponential fashion and stops abruptly as environmental resistance or any other limit becomes effective almost suddenly. Thus in this form, the relatively unrestricted growth is suddenly halted when the population runs out of some resource (such as food or space), or when stresses like release of pollutants, sudden change in salinity due to cause like downpour etc. interfere, or when the reproductive season suddenly terminates due to cause like development of diapause. The upper limit in J-shaped growth pattern is designated by 'N' and when this upper limit is reached, the density remains at this level for a definite time period, or, as is often the case, an immediate decline occurs, resulting in the production of a "*relaxation oscillation*" pattern in density. These events can bc best explained by considering the example of algal blooms in the culture ponds of the shrimp culture farms. Due to addition of nutrients in water, the density of algal population starts increasing at an enormous rate, but as soon as the transparency value of the pond decreases than the optimum requirement for shrimp culture, lime is added to destroy the algal population rapidly and this phenomenon is commonly termed as "algal crash" in aquacultural sphere.

The J-shaped growth form of population is represented by the following simple model:

$$\frac{dN}{dt} = r_m N \qquad \ldots(1)$$

Where, $\frac{dN}{dt}$ = rate of population growth, i.e., change in number per unit time, r_m = capacity for increase in a particular environmental conditions, and N = population size (number). The integrated form of eqn. (1) obtained by calculus manipulation is

$$N_t = N_0 e^{r_m} t \qquad ...(2)$$

Where N_0 represents the population size (number) at time zero, N_t reflects the number at time t and e represents the base of natural logarithms.

The derivation of eqn. (2) from eqn. (1) is given below:

$$\frac{dN}{dt} = r_m N$$

$$or, \frac{dN}{N} = r_m dt$$

$$or, \int \frac{dN}{N} = r_m \int dt$$

$$or, \ln N = r_m t + C$$

$$when\ t = 0,\ then\ N = N_0$$

$$So, \ln N_0 = C$$

Putting the value of C in $\ln N = r_m\ t + C$, the eqn. takes the following form:

$$or, \ln N = r_m t + \ln N_0$$

$$or, \ln N - \ln N_0 = r_m t$$

$$or, \ln \frac{N}{N_0} = r_m t$$

$$or, \frac{N}{N_0} = e^{r_m t}$$

$$or, N = N_0 e^{r_m t}$$

Thus, at time 't', the population size (number) is represented by

$$N_t = N_0 e^{r_m t}$$

S-shaped growth form

Population of any species does not exhibit continuous geometric increase. When a population is growing in a limited space, the density gradually rises until eventually the presence of other organisms reduces the fertility and longevity of the population resulting in the cease of population growth. The growth curve represented by such a population is sigmoid or S-shaped. In sigmoid form, the population increases slowly at first (positive acceleration phase), then more rapidly (logarithmic phase) and finally slows down (negative acceleration phase) with the increase of environmental resistance until a more or less equilibrium level is reached and maintained. This form may be represented by the simple logistic model:

$$\frac{dN}{dt} = r_m N \frac{K - N}{K} \qquad \ldots(1)$$

Where, $\frac{dN}{dt}$ = rate of population growth, i.e., change in number per unit time, r_m = capacity for increase in a particular enviromental condition, N = population size (number) and K = upper asymptote of the sigmoid curve and is referred to as the carrying capacity of the environment.

The equation may be written in two more additional forms as follows:

$$\frac{dN}{dt} = r_m N - \frac{r_m}{K} N^2 \qquad ...(2)$$

or

$$\frac{dN}{dt} = r_m N\left(1 - \frac{N}{K}\right) \qquad ...(3)$$

The integrated form of eqn. (1) is

$$N_t = \frac{K}{1 + e^{a - r_m t}} \qquad ...(4)$$

Where N_t represents the population size (number) at time *t*, *K* represents the maximum size possible or upper asymptote, r_m stands for the capacity for increase in a particular environmental condition, *e* signifies the base of natural logarithms and *a* stands for the constant of integration defining the position of the curve relative to the origin. The derivation of eqn. (4) from eqn. (1) is given below:

$$\frac{dN}{dt} = r_m N\left(\frac{K - N}{K}\right)$$

$$\text{or, } \frac{dN}{N\left(\frac{K-N}{K}\right)} = r_m dt$$

$$\text{or, } \frac{dNK}{N(K-N)} = r_m dt$$

$$\text{or, } \int \frac{dNK}{N(K-N)} = r_m \int dt$$

$$\text{or, } \ln N - \ln(K-N) = r_m t + C$$

$$\text{or, } \ln \frac{N}{K-N} = r_m t + C$$

$$\text{or, } \frac{N}{K-N} = e^{r_m t} e^{c}$$

$$\text{or, } \frac{K-N}{N} = e^{-r_m t} e^{-c}$$

If $-c=a$, then,

$$\frac{K-N}{N} = e^{-r_m t} e^{a}$$

$$\text{or, } \frac{K}{N} - 1 = e^{a-r_m t}$$

$$\text{or, } \frac{K}{N} = 1 + e^{a-r_m t}$$

$$\text{or, } N = \frac{K}{1+e^{a-r_m t}}$$

Thus at time t, the population size (number) is represented by,

$$N_t = \frac{K}{1+e^{a-r_m t}} \quad \text{...(5)}$$

An indepth analysis of the S-shaped growth form reveals that the curve is the net result of the resistance of environmental factors and increment of population density. This is in contrast to the model of J-shaped growth form in which the environmental resistance was delayed until near the end of the increase (Odum, 1971).

3.3. INTERACTION BETWEEN ORGANISMS

Interaction between organisms in an environment is one of the key points behind their successful survival and evolution. Populations of one species change in size and evolve in response to the dynamics and evolution of population of other species. Organisms of one population in a particular environmental system usually interact with others in diverse ways and also in different magnitudes. Population interactions may have positive effect and negative effect and are, in a sense, the foundation stone of community ecology.

Any population interactions may have two aspects: co-dynamic and coevolutionary. When the population interaction involves reciprocal influences on population size, it is referred to as **codynamic**. If the population interaction occurs to such a magnitude that the evolution of the interacting species is affected, then it is **coevolutionary** as well. In addition, when a limited resource is used by two or more individuals, they are in a state of competition. This kind of interaction is called **exploitative.** When competition takes place in the form of direct conflict, it is known as **interference competition** (Park, 1954, 1962; Miller, 1967; Schoener, 1973, 1974).

Theoretically, the interactions between the populations of two species may be symbolized through various types of combinations like 0 0, - -, + +, + 0, - 0, + - (Table 3.1)

These nine types of interactions between organisms can be divided into two broad types, namely, the **negative interactions** and the **positive interactions.** Two principles regarding these categories are especially worthy of emphasis: (1) In the evolution and development of ecosystems negative interactions tend to be minimized in favour of positive symbiosis that enhances the survival of the interacting species, (2) Recent or new associations are more likely to develop severe negative coactions than are older associations.

A very interesting aspect in interaction biology is its dynamicity, that is, the interaction type (either + + or - - or + - etc.) between the organisms of a given species pair is not constant, rather it may change with respect to time under different conditions or during successive stages in their life histories. Thus, two species might exhibit parasitism at one time, commensalism at another, and even completely neutral at another point of time.

Table 3.1: Representation of two-species population interactions.

Sl. No.	Type of Interaction	Species 1.	Species 2.	General Nature of Interaction
1.	Neutralism	0	0	Neither population affects the other
2.	Competition: Direct Interference type	-	-	Direct inhibition of each species by the other
3.	Competition: Resource use type	-	-	Indirect inhibition when the common resource is in short supply
4.	Ammensalism	-	0	Inhibition of population 1; population 2 not affected
5.	Parasitism	+	-	Population 1, the parasite (gainer), generally smaller than 2, the host (looser)

6.	Predation	+	-	Population 1, the predator (gainer), generally larger than population 2, the prey (looser)
7.	Commensalism	+	0	Population 1, the commensal, benefits while population 2, the host is not affected
8.	Protocooperation	+	+	Interaction favourable to both, but not obligatory
9.	Mutualism	+	+	Interaction favourable to both and obligatory

'+' means beneficial effect, '-' means harmful effect and '0' means no effect.

Chapter 4:

Community Ecology

4.1. BIOTIC COMMUNITY CONCEPT

A biotic community may be defined as an assemblage of populations occupying a particular habitat. Communities not only have a definite functional unit with characteristics trophic structures and patterns of energy flow, but they also have compositional unity in that there is a certain probability of clustering of few species together.

The community concept has a great role in scientific management strategy because *"as the community goes, so goes the organism"*. This clearly indicates that the individual entity of an organism is masked with the characteristics of the community to which it belongs. Mosquitoes, the carriers of malarial parasite can be easily controlled by introducing guppy fish (*Poecilia reticulata*) in standing waterbodies who feed on the mosquito larvae. Again the algal bloom in a coastal lagoon can be managed by introducing oysters like *Saccostrea cucullata,* who depend on micro algae for their nutrition. Thus the best way to control and manage a particular organism, whether we wish to encourage and discourage it, is to modify the community, rather than to make a "direct attack" on the organism.

The question of whether a community is really a fundamental unit of organization has a long history of debate in ecology. One view is exemplified by the statement of the British animal ecologist Charles Elton who wrote in 1933 that "....in any fairly limited area only a fraction of the forms that could theoretically do so actually form a community at one time.... The animal community really is an organized community in that it apparently has **limited membership**...".

An alternative view is that a community is a haphazard and unstructured collection of individuals. In this view, the basic properties shared by the individuals of a community are **(1)** the probability of dispersal happened to fetch them to the same place at the same time and **(2)** that the probability of environmental variations supported them to survive at the same time. This view was supported by Henry Gleason (1926), an American plant ecologist, who wrote "... the vegetation of an area is merely the resultant of two factors, the fluctuating and fortuitous immigration of plants and an equally fluctuating and variable environment...".

Although this debate continues even today, but it is a fact that none of the view is cent percent correct. In some pieces of communities, the membership is limited primarily by population interactions, particularly interspecific competition. In other types, interspecific competition is effectively absent, and predation is paramount. In still others, both competition and predation are equally important, and in some, mutualistic interactions are important. Thus there are many probabilities or possibilities in a community ecology.

One important approach to investigate whether a

community is an organized collection of interacting populations or a haphazard collection of individuals is purely based on statistics. Usually two statistical methods are employed: **(1)** association analysis and **(2)** ordination.

Association analysis investigates the extent to which species co-occur. The quardrants are placed throughout an area, and lists of species for each quadrant are recorded. The simplest test for co-occurrence is revealed through a contingency table, which detects an association between variables.

Ordination investigates the distribution of species along environmental gradients. Quadrants are placed at regular intervals along a transect that traverses an environmental gradient such as elevation, soil characteristics or moisture, and the abundance, or biomass, of each species is measured for each quadrant.

4.2. CHARACTERISTICS OF COMMUNITY

Every community has some individual features of its own, which in no way resemble the characteristics of its individual species, but get exposed only with reference to the community level of integration. Five important characteristics of community that we can measure and study are listed here.

Species diversity

This is a measure of the richness of species in a particular habitat, whose value increases as the habitat becomes favourable for existence.

Growth form and structure

A community type can be described by major categories of growth forms like herbs, shrubs, trees etc. or by further detailing of growth forms into categories like broad-leaved trees and needle-leaved trees. These different growth forms determine the stratification, or vertical layering of the community.

Dominance

Depending on the adaptive elasticity of species, very few species dominate in a community and exert a major controlling influence by virtue of their size, numbers or activities. Dominant species are those which are highly successful ecologically and which determine to a considerable extent the conditions under which the associated species must survive.

Relative abundance

It mainly highlights the relative proportions of different species in the community considering the total number of individuals of all the species and the total number of species in the community.

Trophic structure

It reflects the positions of different species in the food webs spun among the members of the community. It is thus helpful in determining the magnitude and direction of the flow of materials and energy from producer to consumer levels in the community.

Self sufficiency

Nutritionally, each community is a group of autotrophic plants as well as heterotrophic animals, and occurs as a self sufficient and perfectly balanced assemblage of organisms.

4.3. ANALYSIS OF COMMUNITY STRUCTURE

The nature and function of a whole community are not determined by all the organisms or the species present in the community, but a relatively few species or species group generally exert the major controlling influence by virtue of their size, number, production efficiency or other activities. Within these species groups which largely control the community energetics and which have profound influence on the survival strategy of other species are known as **ecological dominants.** The degree to which dominance is concentrated in one, several or many species can be expressed through **index of dominance** that reflects the importance of a species or a group of species in the community as a whole. However, if the environmental variables of the ecosystem or the habitat is highly favourable for the survival of all types of species, the value of index of dominance is eclipsed by the values of richness index, evenness index and Shannon Weaner index of general diversity, which express the uniform importance of almost all the species present in the community.

Few useful indices for analysing the community are discussed here.

1. Index of Dominance

In land communities, spermatophytes usually are major

dominants not only among the autotrophs, but in the community as a whole because they provide shelter for the bulk of the organisms in the community and modify physical factors in various ways. In fact the term dominant has been largely used by plant ecologists to mean the "overstory" or tallest plants in the community. Clements and Shelford (1939) have pointed out that animals (consumers) may also exert a controlling force to the communities.

The index of dominance (Simpson, 1949) gives us an idea regarding the magnitude of stress in a particular. It is calculated according to the following expression:

$$\text{Index of domiance (c)} = \sum\left(\frac{n_i}{N}\right)^2$$

Where, n_i = importance value for each species (number of individual, biomass, production etc.) and N = total of importance values.

The index of dominance is always lower where the dominancy is shared by a large number of species (Whittaker, 1965) or the total populations of the community is **uniformly** distributed different species that is usually witnessed in clean, pollution free waters (Osborne *et al.*, 1976) or in habitat having a unique physico-chemical variables. In fact, it is a general phenomenon that when the environmental stress in any particular habitat increases then the community in that particular habitat would be dominated by a fewer species (species that are more resistant to the existing set of physico-chemical and biological variables) which are often referred to as the **opportunistic**

species. The magnitude of the index of dominance is usually high in stressful habitat.

2. Index of Similarity

Two communities, geographically wide apart from each may also show similarity in respect of their species composition. The similarity arises due to the closeness of the two habitats in respect of environmental characteristics.

The index of similarity between two sampling spots is calculated according to the following expression:

$$\text{Index of similarity (s)} = \frac{2C}{A+B}$$

Where, A = number of species present at sampling station "A", B = number of species present in sampling station "B" and C = number of species common to both the sampling stations.

3. Richness Index

The mathematical expressions of richness index are given below:

$$\text{Richness index } (d_1) = \frac{S-1}{\log_e N} \quad \text{(After Margalef, 1958)}$$

and

$$\text{Richness index } (d_2) = \frac{S}{\sqrt{N}} \quad \text{(After Menhinick, 1964)}$$

Where, S = number of species and N = total number of individuals of all the species.

The index reflects the suitability of a particular habitat for successful thriving and growth of different species. The value of the index usually decreases, when environment becomes unfavourable or stressful due to intrusion of some foreign matter e.g., pollutants or any biological organisms that may be a parasite or predator to the existing group of species.

4. Shannon Weaner Species Diversity Index

It is the most commonly used index in ecological studies owing to the fact that it is dimensionless, independent of sample size and expresses the worth of each species (Kochsiek *et al.,* 1971). It is expressed as follows:

$$\text{Species diversity index } (\overline{H}) = -\sum_{i=1}^{S} P_i \log_e P_i$$

$$\text{or, } (\overline{H}) = -\sum_{i=1}^{S} \frac{n_i}{N} \log_e \frac{n_i}{N}$$

where, P_i = importance probability for each species, n_i = importance value for each species and N = total of importance values.

The value of $\overline{H}$ has several ecological explanations. As Margalef (1986) stated that, "the ecologists find in any measure of diversity an expression of the possibilities of constructing feedback systems". Higher diversity, then, signifies longer food chains and more cases of symbiosis (mutualism, commensalism etc.) and greater probabilities for negative feedback control, which reduces oscillations and hence increases stability.

The value of $\overline{H}$ is also a unique indicator of environmental stress. As the sensitive species gradually shift or get eliminated from a habitat with the increase of the magnitude of environmental stress, therefore the species diversity index has been claimed as an effective statistics for predicting the change in environment (Wilhm and Dorris, 1968; Cairns and Dickson, 1971). Thus decrease in the value of $\overline{H}$ indicates an increase in the magnitude of environmental stress on the species, and the gradual restoration of the environmental quality is indicated by an increment in the value of $\overline{H}$ with the passage of time through the recruitment of new species.

5. Evenness Index

It is a measure of the uniformity of different species in a community and the value increases as the environment becomes favourable. The following expression is used to calculate the index:

$$\text{Evenness index (e)} = \frac{\overline{H}}{\log_e S}$$

Where $\overline{H}$ = Shannon Weaner index and S = number of species.

Chapter 5:

BIODIVERSITY

5.1. DEFINITION AND CONCEPT

The word "**Biodiversity**" is a comparatively new entrant in the lexicon of terminologies used in environmental science. However, the term is gaining rapid momentum due to its intricate relationship with human benefit. Although it was a myth few decades back, but today man has started to extract a large variety of products from the living world with the support of powerful technology. Few dreams of man which have recently translated into practical are listed here:

[A] Micro-organisms are today widely used for biological decolourization of triphenylmethane dyes (ethyl violet, methyl green, crystal violet, malachine green, basic fuchsin etc.). Yatome *et al.* (1981) reported the degradation of triphenylmethane dye by *Pseudomonas pseudomalli.* Kwasniewska (1985) demonstrated that oxidative red yeasts *Rhodotorulae* sp. and *Rhodotorulae rubra* had a high biodegradation potential against the crystal violet. Sani and Banerjee (1999) reported that *Kurthia* sp. has higher rates of decolourization of crystal violet in comparison to other micro-organisms. This property of the micro-organisms can be successfully applied to depollute the dye waste water, released from the textile dyeing units and dye-stuff manufacturing industries.

[B] *Azolla pinnata* is a common aquatic weed of small size containing a very useful nitrogen fixing algae *Anabaena azolla* within its leaves. This weed is already used as a biofertilizer in water-logged rice fields. Its prolific growth (333 t/ha/yr), ease of harvesting and possession of high amount of carbohydrates, proteins and fats, coupled with negligible lignin content, make it a good source for the biomethanation process. Apart from the source of biogas generation, the digested *Azolla* slurry plays an important role in fish growth either by triggering the process of plankton production or directly as a partial substitute for conventional fish feed (Das *et al.,* 1994).

[C] Medicinal and aromatic plants constitute a major source of natural organic compounds widely used in medicines (Dobhal, 1994). The antibacterial activity of the extracts from *Acorus calamus, Zingiber officinale, Cinnamomum zeylanicum, Moringa olifera, Ocimum sanctum* and *Piper betle* against *Staphylococcus eureue, Escherichia coli, Pseudomonas aeruginosa* and *Klebsiella* sp. is also a novel gift from the biological resources (Rajendhran *et al.,* 1998).

[D] Seaweeds or the macroalgae is one of the commercially important living renewable resources as they are the source for the production phytochemicals such as agar, carrageenan and algin which are widely used in several industries mostly as gelling, stabilizing and thickening agents. They are also used as human food, animal feed and fertilizer and also possess medicinally active compounds (Table 5.1).

Table 5.1: Activity chart of some major Indian seaweed extracts

Activity / Species	Hypotensive	Toxic	Antiviral	Diuretic	Spasmolytic	CNS depressant	Antifertlity
Caulerpa racemosa	+						
Caulerpa sertularioides	+						
Chaetomorpha media	+						
Cladophora pinnulata		+					
Codium elongaum			+				
Enteromorpha intestinalis			+				
Enteromorpha sp.				+			
Ulva fasicata				+			
Chonospora implcxa				+			
Dictyota bartayresiana				+			
Sargassam tenerrimum							

Spatoglossum asperum				+	
Padina tetrastromatica			+		
Steechospermum marginatum			+		
Acanthophora spicifera					+
Chondria armata	+				
Gelidiella sp.	+				
Gracilaria corticata	+				
Hypnea musciformis		+			

[E] The water soluble part of alcoholic extracts of the leaves of *Azadirachta indica* (Neem) can lower the serum cholesterol level in rats significantly without altering serum protein, blood urea and uric acid (Chattopadhyay *et al.*, 1993). Various parts of this plant have been reported to have medicinal properties like antidiabetic, antiseptic, healing of wounds, curing of skin diseases and antiulcer activities (Chopra *et al.*, 1956).

[F] The mangrove vegetations have now become a source of numerous medicines. Skin disorders and sores, including leprosy, may be treated with ashes or bark infusions of certain mangrove species. Headaches, rhematism, snakebites, boils, ulcers,diarrhoea, haemorhages etc. are traditionally treated with mangrove plants (Sivakumaran and Sanjeevi, 1996).

Certain mangrove species, notably the cedar mangrove and the cannonball mangrove (relatives of the red cedar) as well as the grey mangrove are prized for their hard wood and used for boat building and cabinet timber as well as for tools like spears, digging sticks etc. The fronds of the Nipa palm are used for thatching and basket weaving. The species *Sonneratia apetala* is currently used as bioindicators of dissolved heavy metals in coastal water (Niyogi *et al.,* 1997).

[G] Levin and Bang (1964) suggested the use of Limulus Amebocyte Lysate (LAL) for detection of bacterial endotoxins in injectable pharmaceutical products, food, drugs, vaccines, radio-isotopes, medical devices, water supplies etc. Today it has been established that Carcinoscorpius Amebocyte Lysate (CAL) and Tachypleus Amebocyte Lysate (TAL) have similar biochemical properties and sensitivity which are derived from the cell lysate of the hemolymph of horseshoe crab species found in the Indian subcontinent namely *Carcinoscorpius rotundicauda* and *Tachypleus gigas*.

Among thousands of benefits obtained from the faunal and floral resources of the biosphere, the few countable examples relate not only our intricate relationship with the biotic community for our own sustenance, but also our unique diversification in the sphere of product extraction from the living resources of the globe. Starting from tiny microbes to largest blue whale, human beings are dependent on every chips of life either directly or through indirect routes. The roots of dependency have become so deep, that systematic and rational exploitation of biotic resources have become inevitable. On this background, the term "biodiversity" got a global importance and biodiversity conservation became an international slogan, especially after the signing of the convention on Biological Diversity, in Rio, in mid 1992.

Biodiversity encompasses all species of plants, animals and micro-organisms and the ecosystem and ecological processes of which they are parts. At the same time biodiversity is a global asset, bringing benefits to people in all parts of the world. The term "biodiversity" has been described in a number of alternate ways. It refers to "the wealth of life on the earth, the millions of plants, animals and micro-organisms, the genes they contain and the intricate ecosystems they help to build into the living environment". It is also described as "the varieties of living organisms at all levels, genetic variants belonging to the same species, family and genera, population, community, habitat and even ecosystems". Different classification (by species, communities or ecosystems for example), have been suggested as possible basis for studying biodiversity. These classifications correspond to three fundamental and hierarchial-related levels of biological organization:

Genetic diversity

This constitutes all probable varieties of genetic informations contained in the genes of individuals of plants, animals and micro-organisms. The significance of genetic diversity is often highlighted with reference to global agriculture, food security and biological wars. Genetic diversity arises because of gene and chromosome mutations and the geographic isolation of different populations. It is like an insurance policy to deal with changing conditions in the environment. The characteristics of certain individuals may better equip them to deal with an unexpected change and thus pass on their genes. Since changes in our environment is inevitable, genetic diversity helps guarantee the ability of species to adapt to these changes. Genetic variation within species also provides the raw material from which new species can evolve over time.

Species diversity

This is a function of the distribution and abundance of species. Often, species richness is used almost synonymously with species diversity. However, technically, species diversity includes some consideration of evenness (uniformity) of species abundances.

Ecosystem diversity

This relates to the variety of habitats, biotic communities and ecological processes in the biosphere as well as the diversity within ecosystems. This manner of viewing biodiversity has assumed some significance in particular because a number of environmentalists and ecologists emphasize conservation of biodiversity along with its associated habitats.

5.2. BIOLOGICAL VALUATION OF BIODIVERSITY

Valuation of biodiversity is a complicated task and can be approached through several angles. However, few biological criteria to evaluate a habitat are listed below:

Richness

It is an indicator of the magnitude of variation of species in a particular habitat. The greater the number of species, the higher the biodiversity value of the area.

Rarity

Rarity is related to a notion of scarcity value. The scarcity can, however, emerge either from use value or from existence value.

Distinctiveness

The value concept that is relevant would depend on this. The distinctiveness may be due to rarity or due to use by human beings or any other species.

Representativeness

This criteria can be reflected in the selection of the sample in the context of biodiversity conservation over large areas.

Threat

If any species is overused for serving mankind, it becomes endangered. Rarity results from this overuse. Hence species that are threatened need to be preserved. The biodiversity valuation of an ecosystem falls if the magnitude of threat to the ecosystem (either due to pollution or habitat destruction) increases.

Function

This concept is developed around the notion of the "keystone" species which has a great role in maintaining community structure, composition or process.

Biological criteria, stated above, refer to conservation

for future use or for continued existence as a component of ecosystem. A prioritization based on them reflects the objective function of a society which leaves the ecosystem undepleted in the process of consumption. In other words, human society achieves more benefit and welfare from future cosumption and the maintenance of well integrated ecosystems than it does from present consumption.

5.3. BIODIVERSITY OF MANGROVE ECOSYSTEM WITH SPECIAL REFERENCE TO INDIAN SUNDARBANS

With a coastline of 7515 Kms., exclusive economic zone of 2014.90 thousand sq.Kms. and a shelf area of 452.10 thousand sq.Kms., India has occupied a unique position in the scenario of coastal biodiversity. The eastern, western and southern coast of the peninsular India is ornamented by majestic rivers having extensive and highly productive estuarine areas. West Bengal, a maritime state of the north eastern part of the country, adjacent to Bangladesh, is indented at its southern part by numerous river openings. The important rivers from east to west are Harinbhanga, Gosaba, Matla, Thakuran, Saptamukhi, Muriganga and Hugli which ultimately end up at the Bay of Bengal, but on the way have encompassed about 54 islands criss-crossed and intersected by various creeks and delta distributaries. This deltaic complex at the apex of the Bay of Bengal is the Indian Sundarbans. With a total land and water area of about 10,00,000 ha., the entire Sundarbans ecosystem of India and Bangladesh supports the world's largest mangrove block which is a well known ecosystem of the Tropics. Almost 62 percent of the Sundarbans is situated in Bangladesh while the remaining 38 percent in the western sector, lies within India.

The Indian Sundarbans at the apex of the Bay of Bengal (between 21°40′ to 22°40′ North latitude and 88°85′ to 89°00′ East longitude) is located on the southern fringe of the state of West Bengal covering the major portions of the North and South 24 Paraganas districts. The region is bounded by Bangladesh in the east, the Hugli river in the west, 'Dampier and Hodges Line' in the north and the Bay of Bengal in the south.

With a considerable degree of marine characteristics in major portion of the ecosystem, the important morphotypes of deltaic Sundarbans are beaches, mudflats, coastal dunes, sand-flats, estuaries, creeks, inlets and mangrove swamps. The mangrove forests of the Indian Sundarbans have been variously estimated to cover 4,18,888 hectares (Sidhu, 1963 and UNEP, 1985), 2,00,000 - 3,00,000 hectares (IUCN, 1989) and 4,26,300 hectares by the West Bengal Forest Department. It has further been estimated that approximately 1,78,100 hectares are water areas. The discrepancy in the figures for forest cover possibly arises from the fact that some denote areas designated as forest land which includes both the water bodies and degraded forests, while others represent an assessment of the area covered with vegetations only. The present investigation encompasses an area of about 9630 sq.Kms. of the Indian part of Sundarbans which includes both forest and reclaimed areas as outlined by Sundarban Development Board. This area has recently been declared under "Sundarban Biosphere Reserve" and has a number of fragments like Tiger Reserve, National Park, Wildlife Sanctuaries etc. (Table 5.2).

Table 5.2: Demarcated areas under Sundarbans Biosphere Reserve (After Banerjee, 1965)

Zone	Area (Sq. Km.)
Tiger Reserve	2585
National Park	1330
Water area	1781
Sand char without forest	153
Actual forest coverage	2333

A total of 69 floral species included within 15 families and 29 genera have been recognised in the mangrove forests of Indian Sundarbans, out of which 34 species are true mangrove type (Table 5.3).

Table 5.3: Major mangrove macroflora in Sundarbans ecosystem

Family	Mangrove flora	Type of plant
Acanthaceae	*Acanthus ilicifolius*	Shrub
	Acanthus volubilis	Herb
Aizoaceae	*Sesuvium portulacastrum*	Herb
Amaranthaceae	*Aerva lanata*	Herb
Apocynaceae	*Cebera odollam*	Tree
	Nerium indicum	Shrub
	Thevetia peruviana	Shrub
Arecaceae	*Nipa fruticans*	Tree
	Phoenix paludosa	Shrub

Asclepiadaceae	*Hoya parasitica*	Herb
	Sareobolus carinatus	Herb
Aviceniaceae	*Avicennia alba*	Tree
	Avicennia marina	Tree
	Avicennia officinalis	Tree
Bignoniaceae	*Dolichandrone spathaceae*	Tree
Boraginaceae	*Heliotropium curassavicum*	Herb
Chenopodiaceae	*Salicornia brachiata*	Herb
	Suaeda maritima	Herb
	Suaeda nudiflora	Herb
Combretaceae	*Lumnitzera racemosa*	Tree
Convolvulaceae	*Ipomoea pes-carpae*	Herb
Cyperaceae	*Cyperus exaltatus*	Sedge
	Fimbristylis halophila	Sedge
	Fimbristylis sub-bispicata	Sedge
	Scirpus triquetra	Sedge
Euphorbiaceae	*Excoecaria agallocha*	Tree
	Excoecaria bicolor	Tree
Fabaceae & Caesalpiniaceae	*Canavalia cathartica*	Twinner
	Caesalpinia crista	Shrub
	Cynometra ramiflora	Tree
	Dalbergia spinosa	Tree
	Derris scandens	Shrub
	Derris trifoliata	Shrub
	Derris umbellatum	Shrub
Loranthaceae	*Viscum orientale*	Herb
Lythraceae	*Ammannia baccifera*	Herb
Malvaceae	*Hibiscus tiliaceus*	Tree

	Hibiscus tortuosus	Tree
	Thespesia populnea	Tree
Meliaceae	*Amoora cuculata*	Tree
	Xylocarpus granatum	Tree
	Xylocarpus moluccensis	Tree
Myrsinaceae	*Aegiceras corniculatum*	Shrub
Pandanaceae	*Pandanus faccicularis*	Shrub
Plumbaginaceae	*Aegialitis rotundifolia*	Tree
Poaceae	*Aeluropus lagopoides*	Grass
	Lersia hexandra	Grass
	Phragmites karka	Grass
	Porterasia coactata	Grass
	Pancium repens	Grass
	Paspalum vaginatum	Grass
Rhizophoraceae	*Bruguiera cylindrica*	Tree
	Bruguiera gymnorrhiza	Tree
	Bruguiera paviflora	Tree
	Bruguiera hexandra	Tree
	Ceriops decandra	Tree
	Ceriops tagale	Tree
	Kandelia candel	Tree
	Rhizophora apiculata	Tree
	Rhizophora mueronata	Tree

The brackish waters of Sundarbans support huge masses of phytoplankton and macro-benthic algal species which constitute the dominant primary producers of the ecosystem (Table 5.4), that in turn support a diverse array of zooplankton. During the long course of planktonic survey in the Hugli-Matla estuarine system and mangrove back waters during 1985 to 1994, a total of 35 phytoplankton species distributed over 18 genera have been documented. Among all the phytoplanktons, diatoms seem to dominate the ecosystem. The genus *Chaetoceros* contributed the largest number of species and accounted for the bulk of the diatom population during the peak phase of phytoplankton bloom. The genera like *Coscinodiscus, Biddulphia, and Pleurosigma* are commonly found in the ecosystem through out the year.

Table 5.4: Primary producers (benthic macro algae & phytoplanktons) of Sundarbans mangrove ecosystem

Phytoplanton
Skeletonema sp.
Coscinodiscus excentricus
Coscinodiscus jonesiamus
Coscinodiscus perforatus
Coscinodiscus asteranphalus
Coscinodiscus coneinnus
Lauderia annulata
Rhizosolenia stotterfothii
Rhizosolenia robusta
Rhizosolenia setigera
Rhizosolenia alata
Bacteriastrum hyalinum
Bacteriastrum varians
Bacteriastrum cosmosum
Bacteriastrum princeps
Chaetoceros curvisetus

Chaetoceros iibenii
Chaetoceros peruvianus
Chaetoceros lorenzianus
Chaetoceros indicus
Eucampia sp.
Dittylum brightwelli
Ditylum solanum
Biddulphia sinensis
Biddulphia mobiliensis
Hemidiscus cuneiformis
Diotoma vulgare
Thallassionema mitzschioides
Thallassiothrix frauenfeldii
Pleurosigma elongatum
Pleurosigma normanii
Bacillaria paradoxa
Nitzschia seriata

Benthic macro algae

Chlorophyceae	**Rhodophyceae**
Enteromorpha intestinalis	*Catenella sp.*
Enteromorpha prolifera	*Gracilaria verrucosa*
Ulva lacuta	
Rhizoclonium grande	

The zooplankton community comprises a heterogenous assemblage of animals covering many taxonomic groups including copepods, mysids, lucifers, gammarid, amphipods, cladocerae, ostracods, cumacea, hydromedusae, ctenophores and chaetognaths among holoplankters and polychaete larvae, molluscan and echinoderm larvae, crustacean larvae and fish eggs and larvae among meroplankters (Table 5.5). Among the zooplankton, calanoid copepods constitute the major bulk followed by cyclopoids and harpecticoids. In general higher abundance of zooplankton is encountered during the pre-monsoon periods. The main cause behind the high species diversity index of zooplankton during the pre-monsoon months may be attributed to the coincidence of the breeding period of most of the coastal and estuarine fishes in the months of March and April in this part of the sub-continent.

Table 5.5: List of zooplankton found in Sundarbans ecosystem

Copepod
Paracalnus spp.
Acrocalnus spp.
Acartia spp.
Acartia spinicauda
Acartiella sewelli
Acartiella keralensis
Labidocera euchaeta
Labidocera minuta
Labidocera spp.
Pontella andersoni
Pontellopsis herdmani
Eucalanus subcrassus

Eucalanus elongatus
Eucalanus spp.
Pseudodiaptomus annandalei
Pseudodiaptomus hickmeni
Peusdodiaptomus aurivilli
Pseudodiaptomus tollingeri
Pseudodiaptomus masoni
Pseudodiaptomus daughlishi
Pseudodiaptomus binghami
Pseudodiaptomus spp.
Neodiaptomus strigilipes
Euchaeta marina
Euchaeta wolfendeni
Euchaeta tenuis
Euchaeta concinna
Euchaeta spp.
Centropages dorsispenatus
Centropages furcatus
Temora turbinata
Temora discaudata
Tortanus forcipatus
Candacia bradyi
Canthocalanus pauper
Undimula darwini
Oithona spp.
Saphirella indica
Corycasus danae
Corycasus agilis
Corycassus catus
Halicyclops temuispina
Mesocyclops spp.

Cyclops spp.
Oncaea venusta
Microsetalla rosea
Laophonta spp.
Euterpina acutifrons
Tachidius discipes
Cladorostrata brevipoda
Chaetognath
Sagitta bedoti
Sagitta enflata
Mysid
Mesopodpsis orientalis
Sergestid
Lucifer hanseni
Gammarid amphipods *(unidentified)*
Cumacea
Paradiastylis sp.
Cladocera
Evadne spp.
Ostracoda
Philomedes sp.
Hydromedusae *(unidentified)*
Ctenophora
Pleurobrachia globosa
Beroe spp.

Invertebrates are another important components of this ecosystem which provide valuable food resources for humans & other higher animals. The scat analysis of Royal Bengal Tiger (*Panthera tigris tigris*) revealed the presence of estuarine crabs, providing the direct dependency of higher vertebrates on the invertebrate species. In addition to the resident invertebrate fauna of the ecosystem, the area also invites two important fascinating marine creatures, *Carcinoscorpius rotundicauda* & *Tachypleus gigas* during the onset of premonsoon which is their breeding season. During this period of increasing salinity, the living fossils visit this ecosystem in pair along the expanding tidal waters. The macro-benthic invertebrates exhibit a distinct vertical zonation on hard substata like boulders, light houses & mangrove tree trunks in the lower stretch of the Hugli estuary which remain in close association with different types of macro algal species.

The Sundarban mangrove ecosystem is rich in terrestrial, aquatic & avian faunal species. The forests & water bodies provide dwelling places, habitats, breeding sites & roosting ground for a wide range of vertebrate species encompassing about 250 species of fishes, 8 species of amphibians, 57 species of reptiles, 161 species of birds & 40 species of mammals (Table 5.6) many of which are endangered in other prats of the world (Chaudhuri & Choudhury, 1994).

Table 5.6: Vertebrate faunal diversity in deltaic Sundarbans

Class	Genera/Species (Source: Chaudhuri & Choudhury, 1994)	Genera/Species (Source: Mandal & Nandi, 1989)
Pisces	96/250	100/141
Amphibia	04/07	4/8

Reptilia	40/57	40/57
Aves	120/300	122/161
Mammalia	33/40	32/40

The mangrove forest of Sundarbans is the excellent habitat for the largest tiger (*Panthera tigris tigris*) population in the globe. *Heritiera* & *Phoenix* vegetations, which are rarely inundated or periodically inundated form ideal tiger habitats. During the period of extreme high tides, these habitats are submerged & under such circumstances, the tigers are compelled to lead an amphibious life. Thus the tiger population, in this part of the world have the strange adaptations of drinking saline water. Tigers, in this ecosystem have taken a toll about 1,500 human lives over the last twenty years and the victims are mostly the populations of the adjacent islands whose positions are much below the poverty line. These people venture the deep mangrove forests for collecting honey, wood, fishes, crabs, tiger prawn seeds & various miscellaneous mangrove resources & pose stress to the virginity of the forest and its associated biodiversity.

The mangrove species of Sundarbans ecosystem are under serious threat in the present era due to habitat destruction, unwanted human intrusion, conversion of mangrove patches into aquaculture, salinity fluctuation & deterioration of water quality by wastes of industrial & anthropogenic origin. About half of the mangrove areas in West Bengal has been clear felled (Blasco, 1977). *Heritiera fomes* has become extremely rare in Indian Sundarbans & may be considered endangered in this belt, although the species is abundant in Bangladesh (Chaudhuri &

Choudhury, 1994). The species *Nipa fruticans* has also very limited population in the Indian Sundarbans. The faunal species of Sundarbans has also undergone a compositional change due to salinity fluctuation & habitat destruction. About six vertebrates have disappeared from this ecosystem since the last 200 years (Table 5.7) & about 20 species are in the list of endangered species (Table 5.8). A conservation approach has recently been adopted by declaring this ecosystem as **'National Biosphere Reserve'** to ensure long-term sustainability of the species diversity in this ecosystem.

Table 5.7: Extinct species from Sundarbans mangrove ecosystem

Cairina scutulata (White-winged wood duck)
Muntiacus muntjac (Barking deer)
Cervus deruchea (Swamp deer)
Bubalis bubalis (Wild buffalo)
Rhinoceros sondaicus (Javan rhinoceros)
Rhinoceros unicornis (Indian one-horned rhinoceros)

Table 5.8: Endangered species of Sundarbans mangrove ecosystem

Mammals
Felis viveina (Fishing cat)
Panthera tigris (Tiger)
Platanista gangetica (Gangetic dolphin)
Orcella brevirostris (Irrawady dolphin)

Neophocaena phocaenoides (Little porpoise)
Manis pentadactyla (Chinese pangolin)

Birds

Ardea goliath (Giant heron)
Leptoptilos dubius (Adjunct stork)
Haliaeetus leucogastar (White bellied sea eagle)
Pandion haliaetus (Osprey)

Reptiles

Crocodilus porsosus (Saltwater crocodile)
Lepidochalys olivacea (Olive Ridley Turtle)
Batagur baska (Batagur turtle)
Kachuga teeta (Indian tent turtle)
Lissemys punctata (Indian flap shelled turtle)
Troonyx gangeticus (Indian soft shelled turtle)
Troonyx hurun (Peacock marked soft shelled)
Varanus bengalensis (Common Indian monitor)
Varanus flavesceus (Yellow monitor)
Varanus salvator (Water monitor)
Python molurus (Indian rolled python)

5.4. CONSERVATION OF BIODIVERSITY

Conservation of biodiversity is of utmost importance because clean air, clean water, climate, health and the food supply of all the species of the planet earth depend on biodiversity. Human being is also a part of the world's biodiversity picture and therefore to sustain our own generation in a healthy, pollution-free environment, preservation of biodiversity should be given top priority. Today biodiversity conservation has become an international slogan especially after the gathering at Rio de Janerio during 3-4 June, 1992.

Threats to biodiversity include habitat loss, introduced species, pollution, urbanization and industrialization. The accelerating loss of biological diversity is undermining sustainable development opportunities world wide, eroding genes, species and ecosystems that constitute important resources and support systems to mankind. The first step to biodiversity conservation is to identify the real socio-economic causes and impacts of changes in biodiversity, and to develop strategic means to use the various biotic components in such a way so that their decline due to long term use can be avoided, and they can be rationally exploited for human benefit in a sustainable fashion.

However, the concept of biodiversity conservation has not yet achieved a concrete shape mainly due to two principal reasons (Mahajan, 1996):

- The number of species originally present is unknown and
- There is inadequate information about the rates of loss of habitats.

The entire approach of biodiversity conservation can be broadly divided into three major heads: scientific approach, socio-economic approach and policy approach. Although the concept of biodiversity conservation can be viewed from different angles other than these three approaches, but for simplicity, the areas encompassed within these approaches are discussed here.

Scientific Approach

This approach is purely based on high quality scientific research and biotechnology. Few examples of this mode of conservation are discussed here:

1. **Ecosystem restoration** can bring back the lost essence of biodiversity partly if the afforestation programme is taken up on the basis of soil characteristics and ecological succession. In the language of Edward O. Wilson (1996) "......If, for example, a rain forest is completely cut over, there is no way we could put together all the hundreds and thousands of species, even if they could be saved in some manner. To put them together would be a little bit like unscrambling eggs. However, there is a great deal of hope if the lost vegetations are planted sporadically here and there through land formation and careful cultivation methods.....".

   The concept of ecosystem restoration has also been implemented around some shrimp culture farms in coastal West Bengal, where excess nutrients released by the shrimps are assimilated by the algae. The culture of edible oysters (*Saccostrea* sp.) at the same sites have helped to keep the algal population diversity under control. This type of restoration of water quality by employing biological species is an emerging concept in biodiversity conservation.

2. **Molecular genetics** has an important role to play in

biodiversity conservation. The long-term survival of ecosystems, species and populations is only possible if they carry sufficient genetic variability to withstand and adapt to future challenges from their environment. While it is relatively easy for species to lose genetic variation through bottle necks in population size and inbreeding, there is almost zero chance that lost genes will ever **"re-evolve"**. This means that the main task of biodiversity conservation is to prevent the loss of those components of genetic diversity which determine the **"ability to survive"** of a species or population.

Molecular genetics, in this context, can be used:

- to identify parents, offspring and close relatives in captive and wild populations so as to avoid inbreeding depression of rare and endangered species
- to define the genetic units for conservation
- to quantify the genetic variability of historical populations and their living descendants, and assess the rate at which genetic variation is being lost and restructured in populations in fragmented landscapes.

Reintroduction of species in any habitat without a molecular genetics background may some times be harmful. The example of reintroduction of African hunting dog can be cited in this context. The African hunting dog is a unique carnivore which has historically inhabited Savannah-like habitats south of the Sahara desert. Wild dog populations have greatly declined during the last 20 years and no more than several thousands exist now. The populations in the eastern Africa have recently suffered severe population decline. Molecular genetic analyses showed that eastern and southern populations form two genetically distinct clades that warrant separate preservation. Thus plans to reintroduce south African wild dogs to areas of east Africa (where populations have recently declined severely) would not only mix populations with separate evolutionary histories, but would also cause the potential loss of genes that are unique and adaptive in the east African populations. In addition the introduced south African dogs may put the native eastern populations at risk with south African diseases to which they have not developed immunity. Thus planning to conserve a species in a particular habitat also requires a feedback of molecular genetics.

2. **Captive breeding** is a unique approach to conserve endangered species where breeding is done in captive condition by copying the environmental conditions of their natural habitat. Some successful captive breeding programmes implemented in the Sundarbans Biosphere Reserve in coastal West Bengal are listed here:

Crocodile

The estuarine crocodile (*Crocodylus porosus*) is one of the largest saline water reptiles which was indiscriminately killed for the purpose of making various luxury goods from its skin. The magnitude of poaching became so severe, that the population subsequently declined, making the species endangered.

On 1st April, 1975, the Government of India, with the assistance of FAO & UNDP requisitioned the services of an FAO expert and initiated crocodile farming in India. The salt water crocodile scheme was initiated in Sundarbans on

March, 1976 at Bhagabatpur on Lothian Island and efforts were focussed to reduce the high mortality at the egg and newly hatched stages. Today, the salt water crocodile scheme at Sundarbans is one of the principal crocodile breeding centres within India. By 1990, more than 197 crocodiles had been released into the Sundarbans (Chaudhuri & Choudhury, 1994) which is a unique success indicator of the programme.

Olive Ridley Turtles

Marine turtles are protected under schedule 1 of the Wildlife Protection Act. The Olive Ridley Turtle (*Lepidochelys olivacea*), which extends its migration to Sundarbans deltaic region is an endangered species. Previously there was a period when millions of Olive Ridley eggs were harvested immediately after the breeding season. The poaching of mating pairs was a lucrative business during the breeding season (January & February) when thousands of animals were brought to the shore at Digha, West Bengal for selling in the Calcutta market. This reduced their population to a great extent and therefore attempts were made to rear and rehabilitate this species at Bhagabatpur Crocodile Rearing Centre and at Saptamukhi Hatchery by the Department of Forests, Government of West Bengal. The programme showed an encouraging success when out of 117 hatchlings that emerged from artificial nests at Bhagabatpur, 99 healthy hatchlings were released into the ambient water and 18 were segregated for further research works (Banerjee, 1985).

Horseshoe crabs

Horse shoe crabs, a chelicerate arthropod belonging to the

class merostomata is an amazing creature which is considered to be the oldest **"living fossil"**. There are only four species of the horseshoe crab reported in the world among which *Carcinoscorpius rotundicauda* is found in plenty in the deltaic Sundarbans region. Although *Tachypleus gigas,* is also reported from the deltaic Sundarbans, but the frequency of occurance is very poor (3% to 5%) among the total horseshoe crab population. The species is destroyed at random during the fish catch in the neritic zone and wild harvest of tiger prawn seeds. With the vague belief to cure arthritis, the extracts of different organs of the species is also sold in the Calcutta markets. The buyers are mainly the people belonging to backward class of the society. These activities declined the population of the species in recent years in the North East coast of the Bay of Bengal. The species has been found to be a potential source of bioactive substance, the Limulus Amoebocyte Lysate from its blue blood (Chatterjee and Abidi, 1993). The reagent is highly sensitive and useful for the rapid and accurate assay of gram negative bacteria even if they are present in a very minute quantity upto the level of 10^{-10} gm. Hence the LAL has been proved to be a valuable diagnostic reagent in the detection of endotoxins in several pharmaceutical products especially injectables. Considering the immense biomedical value, of the species, a project entitled "Application of biotechnology and molecular biology in the resource exploration and sustainable management of Sundarbans mangroves" has been sanctioned to the S.D. Marine Biological Research Institute at Sagar Island by the Department of Biotechnology under the Ministry of Science and Technology, Government of India on 26th December, 1997 with the aim to develop culture technology for *Carcinoscorpius rotundicauda* & *Tachypleus gigas* suited to

Sundarbans environment. This project has started to investigate the optimum range of various physico-chemical variables (like water temperature, salinity, pH, nitrate, phosphate, silicate, dissolved oxygen, dissolved copper, organic carbon of sediment and acid-leachable copper of sediment) in relation to their event of fertilization and condition value of three different weight groups. The development and standardisation of culture technology of this species may pave the way of manufacturing Carcinoscorpius Amoebocyte Lysate (CAL) and Tachypleus Amoebocyte Lysate (TAL) from the amoebocyte of these species without damaging their natural population.

Socio-economic Approach

Conserving biodiversity becomes extremely difficult in areas where the needs of human beings are intricately tied up with the natural resources. The people of Sundarbans in the Gangetic delta region, for example, penetrate the deep mangrove forests for want of honey, wood etc. through which they earn their livelihood. Preservation of biodiversity from these section of population would only be possible, if they are provided with some alternative source of income and if they are made as a part and parcel of the whole process of conservation programme.

Ecodevelopment is a unique approach in this context. It is a strategy for protecting ecologically valuable areas from the needs and activities of people living in and around such areas. The approach encompasses three important steps:

- Identification, establishment and development of sustainable alternatives to the biomass resources and incomes e.g., oyster culture or seaweed culture could be an alternative source of income to the prawn seed collectors of the north east coast of the Bay of Bengal who take the risk of spending 8-10 hours/day in average to screen the estuarine water and often get attacked by sharks and other marine creatures. The uprooting of mangrove seedlings by the dragging of their nets can be prevented if they are provided with the technology of oyster culture, seaweed culture or crab culture.

- Involvement of the local people living in and around such protected areas into the conservation, planning and management of the area. This will not only channelise some of the financial benefits of conservation to them, but will also give them a sense of responsibility.

- Raising the level of awareness among the local community regarding the valuation of biotic resources both in terms of ecology and economy. Today, when we are just few steps behind the millennium line, it is a good sign that a large section of the globe have become aware of the dependency of human race on other species. This awareness is due to the efforts of several organisations, some radical like the Green Peace, others doggedly persistent like the Narmada Bachao Andolan. Each has its own method with the same aim to aware the individual about his/her role in nature conservation.

  There are various successful ways of raising awareness. WWF has drawn attention to the highly endangered status of the giant panda by adopting it as their logo, the plight of the blue whale has received international attention because of green peace activity resisting whale fleets, Jacques Cousteau and David Attenborough have filmed several endangered species, while Rachel Carson's world renowned book "Silent Spring" brought the issue of the damage caused by pesticides.

Policy Approach

A number of acts and policies have been adopted to conserve biodiversity. In India three main Acts related to environment protection are: the **Wildlife Protection Act** (1972), the **Forest Conservation Act** (1980) and the **Environment Protection Act** (1986), with their various amendments (Government of India, 1992a).

The National Conservation Strategy and Policy Statement on Environment and Development (Government of India, 1992b) provides directives for the integration and internalization of environmental issues in the policies and programmes of various sectors. The National Forest Policy (1988) complements the Forest Conservation Act (1980) by accelerating the participation of local people in forest conservation activities.

In the sphere of marine and estuarine compartment, India has several legislative acts that are relevant to various uses of oceans and coastal zones. The coastal regulation zone (CRZ), notified in 1991 through the Ministry of Environment and Forests, Government of India gives us a guideline of activities (in a zone wise manner) to the users who wants to use this valuable compartment of nature. Under section 3(1) and (2) of the Environment Protection Act (1986) and Rule 5(3)D of the Environment Protection Rules 1986, the CRZ covers the coastal stretches of seas, bays, estuaries, creeks, rivers and backwaters which are influenced by tidal actions in the landward side upto 500 m from the High Tide Line (HTL), and the land between the Low Tide Line (LTL) and the HTL. In general, the HTL is estimated to be the line upto which the water reaches during highest tide in the spring time. In relation to the state of West Bengal, which is credited to have the Indian Sundarbans at the apex of the Bay of Bengal, the CRZ has been classified into three categories with distinct gradation of activities for a more scientific management of the ecosystem (Table 5.9).

Table 5.9: Structure of Coastal Regulation Zone (CRZ) in relation to West Bengal coast

Category	Area	Important consrevation oriented activities
CRZ-I	All the ecologically sensitive and important areas viz., the Sajnekhali, Lothian and Halliday sanctuaries; reserve forests, wildlife habitat and mangrove forests in the Sundarbans; New Moore and Sandhead islands; important breeding sites of fin fishes, shell fishes and other species; important hotspots from the point of biodiversity; the lower long sand area for its aesthetic beauty and biological diversity; historical and	1. Some eco-friendly communication facilities, construction of schools and hospitals will be allowed. 2. Aquaculture will not be permitted in mangrove rich areas. 3. Seed hatchery will be encouraged from where 5% will be released in the creek. 4. A marine park will be established in the Southern part of the Sagar Island.

	heritage sites at Sagar Island that fall within the high tide level (HTL) and low tide level (LTL).	5. Culture of algae and edible oyster will be encouraged.
CRZ-II	Areas that have already been developed and close to the shore line. Such areas include those within the municipal limits (e.g., Haldia along the Hugli river) and other legally designated areas which have drainage facilities, approach roads, water supply etc. The Haldia part and eastern zone of Digha fall within this zone.	1. Shore protection measures will be adopted. 2. Dredging activities will be regulated. 3. EIA will be compulsory for any project and regular monitoring of pollution will be carried out.
CRZ-III	Areas that are yet to be developed and include Shankarpur and Western Digha.	Aquaculture sites may be selected on the basis of recommendation by a special body of State Government.

Chapter 6:

Environmental Pollution

6.1. DEFINITION AND TYPES OF POLLUTANTS

Pollution may be defined as an undesirable change in physical, chemical or biological characteristics of air, water and land that may or will adversely affect human life, industrial progress, living conditions and cultural assets.

The substances that are responsible for the undesirable changes in the atmosphere are referred to as the **pollutants**. Thus, pollutant is a substance (e.g., dust, smoke), chemicals (e.g., SO_2 or Methyl mercury) or factor (like heat, noise etc.) that on release into the environment has an actual or potential adverse effect on human interests. According to the Indian Protection Act (1986) "*....a pollutant has been defined as any solid, liquid or gaseous substance present in such concentration as may be or tend to be injurious to the environment....*". Thus, pollutants can be stated as substances present at a wrong place, in wrong amount and in wrong time.

Few common examples of pollutants are listed in Table 6.1.

Table 6.1: Major Types of Pollutants

Type	Examples
Gaseous pollutants	Oxides of nitrogen, SO_2, H_2S, CO, halogens (chlorine, bromine, iodine etc.)
Fluoride compounds	HF, F_2, SiF_4, H_2SiF_6 etc.
Metals	Hg, Cu, Mn, Cr, Pb, Fe, Zn, Ni, Sn, Cd, etc.
Organic pollutants	Benzene, benzapyrene, toluene, xylene etc.
Photochemical oxidants	Ozone, PAN, aldehyde, ethylene etc.
Deposited matter	Soot, tar, dust, grit
Solid wastes	City garbage
Fertilizers	Lead arsenate
Economic poisons	Herbicides, Fungicides, pesticides, insecticides, rodenticides, etc.
Radioactive wastes	Alpha and beta particles, gamma radiations
Noise	Caused by fighter planes that are used by defence.
Heat	Caused by thermal power plants.

Pollutants can also be classified into two basic types

nondegradable pollutants and **biodegradable pollutants** (Odum, 1971).

Substances like aluminium cans, heavy metals (like Zn, Cu, Mn, Fe, Co, Ni, Pb, Hg etc.), long-chain phenolic chemicals and DDT either do not degrade or degrade at extremely slow rates. Such substances are grouped under the category of non-degradable pollutants. These groups of pollutants often accumulate within the living tissues (a phenomenon called **bioaccumulation**) and get "**biologically magnified**" while they move along the food chains towards the members of higher trophic level.

The biodegradable pollutants include domestic sewage which can be easily decomposed by microbial actions into smaller fragments or elements, that are again utilized by the producers for carrying out their nutrition.

6.2. AIR POLLUTION

Some tragic incidences due to air pollution occurred in the Meuse valley of Belgium in 1930, London in 1952 in which about 4,500 people died as a result of respiratory failure due to deadlier smog, Bhopal (India) in 1984 in which 5,000 people were killed approximately and around 2,00,000 residents were affected in some way or other due to leakage of poisonous MIC gas from the Union Carbide Pesticide plant. Air pollution is not a new kind of problem and is considered as one of the most dangerous and common kind of environmental pollution in almost all the industrialized and urbanized cities and towns of the world.

According to **WHO**, air pollution is defined as "*....Phenomenon in which substances put into air by the* *activity of mankind into concentration sufficient to cause harmful effect to his health, vegetables, property or to interfere with the enjoyment of his property....".*

Causes of Air Pollution

The sources of air pollution are many in numbers. The important ones are discussed here.

1. Industrial pollutants

Common air pollutants which are discharged into air from industrial chimneys and power houses are CO, CO_2, SO_2, H_2S and hydrocarbons. These gases are produced due to burning of coal and petroleum and by combustion of lignite at thermal power stations. Various fluorine compounds (HF, F_2, SiF_4, H_2SiF_6, etc.) are emitted by phosphate fertilizer manufacturing plants (since large amounts of fluorine are present in phosphate rock), aluminium extraction (where cryolite i.e., Na_3AlF_6 is used to lower the melting point of the mixture of CaF_2, Al_2O_3 and Na_3AlF_6), ceramic firing, steel making and some chemical processing.

Many metallurgical processes release dust and fumes which are loaded with Pb, Cr, Ni etc. Many chemical industries release hydrochloric acid, chlorine, oxides of nitrogen, Zn, Pb, As, oxides of Cu, in addition to SO_2, CO and H_2S.

2. Mobile combustion sources

These sources include locomotives, air crafts, automobiles etc. Pollution emissions from aircrafts are small, but these emissions unlike most air pollutants are not concentrated in the urban zones. Aircrafts are responsible for about 2.5%

of the carbon monoxide emissions and about 1% of the hydrocarbon emissions.

Automobiles release CO, oxides of nitrogen and hydrocarbons. It has been found that every gallon of petrol consumed by automobiles produces 3 lbs of CO and 15 lbs of nitrogen oxide which is sufficient to pollute 8,00,000 to 20,00,000 cc of air. The combustion of petroleum emits various particulate lead compounds. The particulate lead compounds emitted from petroleum combustion have profound adverse influence on biotic community. The residents of Mumbai have a lead content of 16-18μg/100ml of blood, while lead begins to show toxic effect beyond 20μg/100ml. The larger lead particles settle down on the soil and pass into food chain. The concentration of lead in urban environment is due to use of tetra-alkyl lead, in petrol. Among lead and tetra-alkyl lead, the later is more poisonous.

Benzene, toluene and xylene are the three volatile organic compounds (VOC) present in urban atmosphere mainly due to use of petrol and diesel in automobiles. Out of these three organic compounds, the concentrations of benzene in air showed strong correlations with the incidence of cancer. The limit value of benzene in air for health safety should not exceed 10μg/m^3.

3. Burning of fuels

Fossil fuels are the sources of energy for cooking, heating, lightening our houses, washing clothes through washing machines, or for running TV etc. Coal on combustion produces CO_2. Incomplete combustion yields CO and a variety of hydrocarbons including methane and soot. Ashes and SO_2 are also the products of coal burning. It has been recorded that electrical power plants burning fossil fuels, particularly coal and sometimes petrol or diesel, produce two-thirds of the SO_2.

In majority of the Indian cities and towns, cooking is done by burning coals, which is an important cause behind the increment of benzene concentration in the atmosphere. While in most of the developed countries, the most toxic benzene is much less than toluene and xylene in the urban air, in the third world countries, the picture is totally reverse. In Calcutta, more than 80% of the families still use coal chullas which account for high benzene load in the city air.

When a coal fed stove (chulla) burns, the concentration of benzene may be 5497μg/m^3 adjacent to the chulla. A comparative study of ambient benzene concentrations in few cities of the world is shown in Table 6.2.

Table 6.2: Average value (during 1994-1995 winter) of ambient benzene concentrations in major cities of the world (After Chakraborti, 1996).

Calcutta	London	Boston	Los Angeles
1500 μg/m^3	40 μg/m^3	60 μg/m^3	140 μg/m^3

4. Agricultural activities

Burning of forest areas, grasslands etc. for pastures and croplands produces about 60 to 65% of CO_2. About 40% of methane is produced from paddy fields, guts of livestocks and also from burning of biomass.

Crop spraying and dusting for pest and weed control are responsible for emitting organic phosphates, chlorinated hydrocarbons, arsenic and lead into air.

5. Ionizing radiations

Ionizing radiations like alpha and beta particles are produced during nuclear explosions, scientific experiments where radioisotopes are used and atomic weapon testing. Radiations are also caused by naturally occurring radioactive substances. Table 6.3 highlights a list of radioactive isotopes, their approximate abundances in the lithosphere, their half lives and the nature of their radiations.

Table 6.3: List of some naturally occurring radioactive substances

Isotope	Abundance	Half-life (years)	Radiation
Ra^{226}	2×10^{-12}	1622	α, γ
U^{238}	4×10^{-6}	4.5×10^{9}	α
Th^{232}	2×10^{-6}	1.4×10^{10}	α, γ
K^{40}	3 ppm	1.3×10^{9}	β, γ
V^{50}	0.2 ppm	5×10^{14}	γ
Rb^{87}	75 ppm	4.7×10^{10}	β
In^{115}	0.1 ppm	6×10^{14}	β
La^{138}	0.01 ppm	1.1×10^{11}	β, γ
Sm^{147}	1 ppm	1.2×10^{11}	α
Lu^{176}	0.01 ppm	2.1×10^{10}	β, γ

These naturally occurring radioactive substances are

important sources of radiation exposure to humans. The amount of whole-body radiation produced annually has been found to vary from place to place, depending on variables such as the nature of the soil and the distance from the source (mineral deposits). In Brazil, there are some habitats where the soils are extremely rich in thorium and uranium. In such places the whole-body exposure of the residents can amount to 2 rem annually.

6. Cosmic radiations

The earth's atmosphere is under the continuous bombardment of high energy particles coming from outer space. These particles are the **primary cosmic rays**. As they penetrate into the atmosphere, primary cosmic ray lose energy and gradually disappear on colliding against the oxygen and nitrogen atoms of the air. At the same time, however, they give rise to secondary rays mostly different in nature from the primary particles and, of course, of lower average energy. Thus, at each point within the atmosphere, one finds a radiation consisting partly of primary particles, and partly of secondary particles. This radiation may be denoted as **local cosmic radiation**.

The primary cosmic radiation consists of protons, alpha particles and to a smaller extent, heavier nuclei. The energy spectrum of protons obeys approximately the following empirical law:

$$N(E) = \frac{A}{(E+5.3)^{1.75}}$$

where, E is the kinetic energy (measured in Bev), N(E)

is the number of protons with energy greater than E, and A is a constant.

When a primary particle of particularly high energy strikes the atmosphere, it initiates a chain of interactions that continues through many successive generations and thus gives rise to very large number of secondary rays. Among these, electrons and photons are the most abundant, but protons, neutrons and mesons are also present. This phenomenon is called an **extensive air shower** (or an **Auger shower**). Table 6.4 lists some products of cosmic rays.

Table 6.4: Products of cosmic radiation, their half-lives, and their concentration in disintegrations per minute per cubic meter of air in the lower troposphere.

Isotope	Half-life	Concentration
H^3	12.3 years	10
C^{14}	5760 years	4
Be^7	53 days	1
S^{35}	87 days	0.015
P^{33}	25 days	0.015
P^{32}	14.3 days	0.02

The radiation dose obtained from cosmic rays has been found to be greater at greater heights in the atmosphere and may amount to about 10 rem/year above the atmosphere. Natural radiation from cosmic rays will be higher at higher elevations and higher in a jet aircraft at 10 km than on the ground.

7. Suspended Particulate Matter (SPM)

SPM is a major air pollutant. Dust is generated from sources such as coal dust (from oil refineries, power plants etc.), cement dust, silica dust (from stone crushing). A huge amount of dust is also blown by transport vehicles.

Incidences of high lead dust were recorded in some pockets of the city of Calcutta. In Picnic Garden area of Calcutta, there are 27 lead factories producing lead ingots and lead alloys. In the dining tables of some local houses, the concentration of lead in the dust exceeded 5,000 μg/gm (Chakraborti, 1996).

Effects of Air Pollution

Dust, smoke and other suspended particulate matter reduce the visibility. Some of the aerosols as H_2SO_4 mist, $(NH_4)_2SO_4$ mist and water vapour influence the vertical temperature profile in the atmosphere. It affects the thermal mixing and also leads to green house effect and depletion of ozone layer. Air pollution also reduces the aesthetic value of places and exerts considerable adverse effects on the health of biota.

The various effects of air pollution are discussed separately.

Effect of Air Pollution on Climate

A large amount of CO_2 gets introduced into the atmosphere due to burning of fossil fuels, furnaces etc. The gas gets confined exclusively to troposphere. In dense concentration, it can act as a serious pollutant. The temperature at the surface of the earth is being maintained by the energy balance between the sun's rays that strike the planet and the heat that gets radiated back into the space. Some of the Sun's rays that penetrate the thick layer of CO_2 are able to strike the earth and get converted into heat. The heated earth is able to re-radiate this absorbed energy as radiations of **longer wavelengths**. Much of this does not pass through CO_2 layer to the outer space but gets absorbed by the CO_2 layer creating an addition of extra heat. An increased heating of earth may result in melting of the polar ice caps and finally cause a rise in the sea level. This phenomenon is termed as **green house effect**. Many workers have predicted that there is a high probability of increment of the global temperature by 1.5 to 4.5°C by the year 2050.

Sometimes, in the presence of sun light, atomic oxygen from the photochemical reduction of NO_2 also reacts with a number of hydrocarbons (such as methane, ethane, toluene etc., all of which originate from burning of fossil fuels or directly from plants) to form reactive intermediates called **radicals**. These radicals then take part in a series of reactions to form still more radicals that combine with oxygen, hydrocarbons and NO_2. This results in the regeneration of NO_2, disappearance of nitric oxide, accumulation of ozone and formation of a number of secondary pollutants like formaldehyde, aldehydes and peroxyacetyl nitrate or PAN ($C_2H_3O_5N$). All these collectively form **Photochemical smog**.

The increased use of CFCs by human beings in recent times has opened another dark chapter in air pollution. These obnoxious compounds emitted mainly by aerosols escape to stratosphere and react with ozone. A sharp drop in the protective ozone layer (about 40%) over Antarctica could have been caused by human produced pollutants. This may ultimately pave the way of global temperature rise.

The effects of major air pollutants are schematically represented below:

CFSs and NO_X → Ozone depletion

CO_2 and CH_4 → Global Warming

SO_2 and NO_X → Acid Rain

PAN and O_3 → Photochemical smog

Effect of Air pollution on Human Health

Human health is seriously affected by air pollution and in extreme cases many diseases may be fatal. Diseases like asthma, bronchitis, pollen allergies, lung cancer are all associated with air pollution. Baker *et al.* (1968) reported that in many American cities along the Eastern Seaboard increasing frequencies of bronchitis, cough, soar throat, wheezes, eye irritations and general ill health in people occurred as air pollution level increased. Mountain *et al.* (1968) reported the occurrence of respiratory problems in children under 8 years in the city of New York.

Air pollution can also be a cause of cancer. Higher

benzene soluble organic matter (BSOM) in city air means that the air is more polluted and usually automobiles and coal burning are responsible for higher BSOM value. In BSOM, there lie a group of compounds known as Polynuclear Aromatic Hydrocarbon (PAH) and many of them are suspected **carcinogens** (like benzapyrene). Evidences from epidemiological studies suggest an increase of one benzapyrene unit corresponds to an increase of 5% in the lung cancer death rate.

The pathological effects of various air pollutants along with their sources are given in Table 6.5.

Effect of Air Pollution on Aesthetic Value

1. Coal dusts are a source of nuisance for household articles and impart a dirty look.

2. The stone in Parthenon in Athens has deteriorated in the past 50 years for air pollution. Similarly statue of Liberty is corroded from SO_2 and NO_2 and Taj Mahal from SO_2 emitted from Mathura refineries. Soot and tar stick to the building stones and painted surfaces, which are very difficult to remove completely.

Prevention and Control of Air Pollution

Air pollution is regarded as a grave danger for the healthy sustenance of all forms of life on this earth. The various measures to control air pollution are given below:

1. Sulphur free and lead free fuel should be used for motor vehicles.

2. The height of the chimneys of factory should be tall

Table 6.5: Pathological effects of some common air pollutants (After Southwick, 1976)

Pollutants	Source	Pathological effect
Aldehydes	Thermal decomposition of fats, oils and glycerol	Irritate nasal and respiratory tracts
Ammonia	Chemical processes, dye making, explosives, fertilizer	Inflames upper respiratory passages
Arsine	Processes involving metals or acids containing arsenic soldering	Breaks down red cells in blood, damages kidneys, causes jaundice
Carbon monoxide	Gasoline motor exhausts, burning of coal	Reduces oxygen carrying capacity of blood
Chlorine	Bleaching cotton and flour, many other chemical processes	Attacks entire respiratory tract and mucous membranes of eyes, causes pulmonary edema
Hydrogen cyanide	Fumigation, blast furnaces, chemical manufacturing, metal plating	Interferes with nerve cells, produces dry throat, indistinct vision and headache

Hydrogen fluoride	Petroleum refining, glass itching, Al and fertilizer production	Irritates and corrodes all body passages
Hydrogen sulphide	Refineries and chemical industries, bituminous fuels	Causes nausea, irritates eyes and throats
Nitrogen oxides	Motor vehicle exhausts, soft coal	Inhibits cilia action so that soot and dust penetrates far into the lungs
Phosgene (Carbonyl chloride)	Chemical and dye manufacturing	Induces coughing irritation and sometimes fatal pulmonary edema
Sulphur-di-oxides	Coal and oil combustion	Causes chest constriction, headache, vomiting and even death from respiratory ailments
Suspended particles	Incinerators, any manufacturing units	Causes emphysema, eye irritations and possibly cancer

enough to reduce the rate of pollution at the ground level.

3. To remove the particulate matter in the smoke, it should be filtered before releasing into the air.

4. Plantation should be done on a large scale. Plants like *Ficus variegata, Phaseolus vulgaris, Coleus bulmeri* can fix CO easily. Some other plants like *Pinus, Juniperus, Quercus, Vitis* can use oxides of nitrogen. To reduce the effect of automobile pollution, the free spaces of the city area may be given a vegetation cover on the basis of the Air Pollution Tolerance Index (APTI) of trees (Table 6.6).

5. Control equipments like gravity settling tanks or porous filters and electrostatic precipitators should be installed in factories to minimize air pollution. The method used for pollutant separation is dependent on the size of the pollutant particles. **Gravity settling chambers** are used for particles having size more than 50μm. **Electrostatic precipitators** or **Cyclone collectors** are used for small sized particles.

   Cyclone collectors contain a chamber in which gas stream with particulate matter is whorled round through a tight circular spiral. Thus particulate pollutants are centrifuged, collected and removed. In electrostatic precipitators, precipitation is due to electric charging of particles and collecting surface bears an opposite charge to attract them. Through this technique, particles with size of 5-20 μm can be removed to the extent of 95%.

6. Permanent air monitoring stations should be set up in

Table 6.6: Evergreen trees arranged in decreasing order of their APTI determined on the basis of leaf extract pH, ascorbic acid, total chlorophyll and relative water contents (RWC)

Plant species	Total chlorophyll (mg/gm dry wt.)	pH	Ascorbic acid (mg/gm dry wt.)	RWC %	APTI
Ficus glomerata	18.89	8.9	9.41	71	32
Pithercelobium dulce	17.21	6.0	8.35	87	28
Polyalthia longifolia	6.56	6.0	8.75	78	19
Ficus infectoria	6.58	7.9	7.51	76	19
Nerium odorum	2.39	6.0	8.17	84	15
Eucalyptus citriodora	4.59	5.2	6.61	85	15
Leucaena leucocephala	10.41	5.7	4.42	72	14
Mangifera indica	3.89	5.2	4.12	72	11

Anona squamosa	4.66	5.4	4.11	60	10
Syzygium jambolana	7.07	4.5	2.86	68	10
Acacia arabica	2.54	6.1	3.94	54	9
Artocarpus heterophyllus	6.30	6.2	3.12	48	9
Alstonia scholaris	3.68	6.5	2.89	64	9
Grewia asiatica	1.85	5.6	2.10	74	9
Casuarina equisetifolia	0.75	5.3	1.79	56	7
Nyctanthes arbortristis	6.96	6.8	1.46	34	5
Litchi chinensis	1.74	6.1	0.70	40	5

the core area of the city to monitor the air quality through out the year. Any deviation from the standard values of different variables of ambient air (Table 6.7) should be analysed properly to know the actual cause either through "**principal component analysis**" or any other package that can highlight both the synergistic or antagonistic effects of variables on the deviation.

Table 6.7: Standards for ambient air quality (in $\mu g/m^3$)

Zone	**SPM**	**SO_2**	**CO**	**NO_X**
Industrial and mixed area	500	120	5000	100
Residential and rural	200	80	2000	80
Sensitive (S)	100	30	1000	30

S = hill station, national parks, monuments

7. Strictly implementation of technology for control of emission from all types of vehicles using petrol or diesel, is extremely important. The Euro-I and Euro-II emission norms need to be mentioned in this context.

   In a very lucid way, Euro-I compliance means that there has to be strict maintenance of the 1:15 fuel:air ratio in all engines so that emission of harmful gases like carbon monoxide and nitrous oxides are kept to the minimum. However, in a carburettor driven engine, as is the case in many domestic cars or in non-Euro-I cars, there is little control over this ratio. Thus the conversion of excess emission of the harmful gases — carbon monoxide and oxides of nitrogen — into less noxious end products is done through catalytic convertors where a coating of noble metals help catalyse to the changes. The only way out of the problem is to employ multi-point fuel injection (MPFI) systems to meet the Euro I norms. For Euro-I, the exhaust is equipped with a computer that studies the emissions and commands the MPFI in such a manner that the ratio is maintained. However, a 16% variation between the approved pollution levels and the spot checks is allowed in Euro-I. But in case of Euro-II, no relaxation is allowed between the approved pollution and spot checks and thus emphasizes a very rigid control on the component quality.

In order to implement these two emission norms, the **Supreme court of India** stated the following clarifications:

There will be no restriction on vehicles running on Compressed Natural Gas.

Only vehicles confirming to Euro-II norms can be registered in the National Capital Region from 1st April, 2000.

The court in its 29th April, 1999 order banned registration of any private non commercial vehicle not meeting Euro-II norms from 1st April, 2000 and allowed registration of 250 diesel and 1259 petrol driven vehicles in the National Capital Region a month on 1st May, 1999 till 31st March, 2000 if they conform to Euro-I emission norms.

6.3. WATER POLLUTION

The National Water Commission stated in 1973 that "....*Water gets polluted if it has been not of sufficiently high quality to be suitable for the highest uses people wish to make of it at present or in the future....*". Infact the term water pollution refers to the addition of substance or heat to such a level, which is harmful to humans, animals or desirable aquatic life, or otherwise causes significant departures from the normal activities of various living communities in or near bodies of water.

Causes of Water Pollution

The major wastes that find their entry into the waterbodies and cause pollution may be grouped into domestic wastes, industrial wastes, agricultural wastes, shipping wastes, radioactive wastes and aquacultural wastes.

1. Domestic wastes

These include waste water from homes and commercial establishments. The domestic wastes are usually charged with nitrates and phosphates and are often responsible for **eutrophication** — a phenomenon of rise of algal density due to enrichment of the ambient water with nutrients. When domestic wastes are dumped into waterbodies, the aquatic community in the immediate vicinity do not show any significant change. This is because the organic wastes do not get time to decay just at the outfall site and therefore the dissolved oxygen level does not show any considerable variation. However, somewhat further away, conditions become worse as bacteria and fungi that degrade organic sewage use up so much oxygen that anaerobic conditions prevail. This is the main reason, why the aquatic phase of DWF and SWF canals that act as the conveyer belt of the municipal and industrial wastes of the city of Calcutta show a very low dissolved oxygen value (about 1.5 mgl^{-1}) in day time inspite of the presence of adequate producers.

2. Industrial wastes

Industries are the major sources of water pollution. A large number of factories and industries located on the banks of rivers and also in the coastal areas discharge their effluents without adequate treatment into the surrounding waterbodies. Most of the Indian rivers are polluted by various types of industries (Table 6.8).

Table 6.8: Sources of pollution of some major Indian rivers

River	Sources of pollution
Kali at Meerut (U.P.)	Sugar mills; distillaries; paint, soap, rayon, silk, tin and glycerine industries.
Jamuna near Delhi	DDT factory, sewage, Indraprastha Power station (Delhi)
Ganga at Kanpur	Jute, chemical, metal and surgical industries; tanneries, textile mills and great bulk of domestic sewage having high organic matter.
Gomti near Lucknow (U.P.)	Paper and pulp mills; sewage.
Dajora in Bareilly (U.P.)	Synthetic rubber factories.
Damodar between Bokaro and Panchet	Fertilizers, fly ash from steel mills, suspended coal particles from

washeries	and thermal power station.
Hugli near Calcutta	Power stations; paper pulp, jute, textiles, chemical mills, paints, varnishes, metal, steel, hydrogenated vegetable oils, rayons, and soap, match, shellac, and polythene industries and sewage.
Sone at Dalmianagar (Bihar)	Cement, pulp and paper mills.
Bhadra (Karnataka)	Pulp, paper and steel industries
Coom, Adyar and Buckinghum canal (Madras)	Domestic sewage, automobile workshops
Cauvery (Tamil Nadu)	Sewage, tanneries, distillaries, paper and rayon mills.
Godavari	Paper mills.
Siwan (Bihar)	Paper, sulphur, cement, sugar mills.
Kulu (between Mumbai and Kalyan)	Chemical factories, rayon mills and tanneries.
Suwao (in Balrampur)	Sugar industries.

Sometimes the industrial wastes carry bulk of conservative wastes (non-degradable pollutants), which accumulate inside the body tissues of flora and fauna inhabiting the system where the industrial wastes are discharged. Such process is highly dangerous as it can pose threat to human health through food chain.

A very recent study conducted in a brackish water farm

at Kanmari, (about 50 km away from the city of Calcutta) where the impounded waterbodies receive the metropoliton wastes of the city and are used for culturing tiger prawns (*Penaeus monodon*), showed significant accumulation of Zn, Cu, Mn, Fe, Cr, Ni, Pb and Cd in the body tissues of the prawns (Mitra *et al.* 1999). These heavy metals find their way into the waterbodies of the area from various industries like painting, electroplating, galvanizing and tannery (Mitra, 1998). The heavy metals carried by the industrial wastes into these waterbodies at the outskirts of the city of Calcutta not only accumulate within the prawn tissues, but also pose a negative stress on their growth (Mitra, 1999).

3. Agricultural Wastes

The agricultural wastes include sediments, fertilizers, and farm animal wastes. These pollutants can enter the waterways as run-off from agricultural lands. Agricultural wastes may also cause ground water pollution. Bioresistant pesticides, which are sometimes heavily applied to deal with large infestations of insects, can perlocate down to ground water. Agricultural wastes can also lead to eutrophication in the waterbodies adjacent to agricultural lands as plant nutrients and fertilizers contribute appreciable amount of nitrates and phosphates to the ambient media.

4. Shipping Wastes

These include both human sewage and other wastes, the most important of which is oil. About half of the crude oil produced per year is transported by sea. After unloading of cargo of oil from a tanker, it carries seawater as ballast. It is a general practice to fill several of the tanks (25% to 30% of the total capacity of the tanker) with seawater to ballast the ship for the voyage back to the loading terminal. These ballast waters are discharged prior to filling the tanks with new oil. When a tanker is unloaded it cannot be completely cleaned because some oil remain on the bottom and some cling to the sides of the tank. The ballast water inevitably become contaminated with these remaining oil and is responsible for much of the causal oil pollution.

Oil pollution also occurs in coastal waters through tanker accidents. These accidents usually occur in narrow straits or near entry points of ports very close to the shore and hence the impact of tanker accidents is witnessed maximum on the planktonic biota of coastal zone. The oil spill in the Hugli estuarine stretch at Haldia in 1996 caused significant lowering of ichthyoplankton diversity (Mitra and Bhattacharyya, 1999).

5. Radioactive Wastes

Coastal waters receive radioactive wastes mainly from the nuclear power plants, nuclear powered ships and submarines. Also extensive use of radio isotopes as tracers in the fields of medicine, biology, biochemistry and biotechnology contribute radioactive substances in the city's drainage system.

Other major sources of radioactive pollutants in the marine and estuarine compartments are those originating from weapon testing and leaking of radioactive material from soil which add to the amount of radiation in the sea. Thus, many radioactive elements like uranium, thorium, actinium etc. occur in the ambient media of the sea. Many of these radioactive elements have a very lengthy half life period and they decay at an extremely slow rate. Such elements tend to bioaccumulate inside the tissues of the biota present in the ecosystem.

6. Aquacultural Wastes

Aquaculture is a growing industry in the present era. Although aquaculture in the freshwater systems have not flourished yet to the expected target, but the coastal aquaculture, particularly the shrimp culture has achieved a unique position in the Indian sub-continent. Although a variety of finfishes, oysters, mussels, sea cucumbers, crabs, and lobsters are in the export list of India, shrimp is the single dominant item in the export basket of our marine food which accounts for almost two-thirds of the total export earnings. To increase the rate of shrimp production (which is usually expressed in tonnes per hectare) from the shrimp culture firms, the stocking density is also greatly increased along with simultaneous input of artificial feed and fertilizers. This greatly increase the nutrient concentrations and organic load per unit area or per cent weight of shrimp production which often exceeds the assimilary capacity of system and results in aquatic pollution. The left over artificial feed supplied for the shrimp growth also generates H_2S in the pond bed.

The major sources of water pollution are listed in table 6.9.

Table 6.9: Major sources of water pollution

Sources	Major pollutants
DOMESTIC	
Biological wastes	Pathogens and biodegradable solids
Detergents	Boron and phosphates
INDUSTRY	
Mining	Lead, zinc, copper and H_2SO_4
Iron and steel plants	Phenol, ammonia, cyanides, cyanates, iron salts
Oil refineries	Free oil, mercaptans, phosphates, phenols, zinc
Paper	Caustic soda, sulphates, chlorine, organochlorides, dioxins
Leather	Chromium, chlorine
Sugar	Nitrogen, biological wastes
Distilleries	Chlorides, potash, sulphates
Thermal power plants	Sr-90, U-235, Pu-239, Cs-137, Ra-226, hotwater, fly ash
AGRICULTURE	
Fertilizers	Free ammonia, phosphates, nitrates, arsenic, fluorides
Pesticides	Organochlorines (BHC, DDT, malathion, parathion etc.)
SHIPPING	

Sewage	Biodegradable solids, nitrates, phosphates, microbes
Oil molecular	Hydrocarbons of varying weight, small quantities of nitrogen, sulphur, vanadium, nickel, zinc, iron
AQUACULTURE	
Feed	H_2S
Excretory products	NO_3, PO_4, unionized NH_3, ammonium ion (NH_4^+)
Antibiotics	Oxytetracycline

In the marine environment, many additional activities like offshore oil exploration, ocean mining etc. are common which often pollute the seawater. The various ways through which the seawater gets polluted are mentioned in table 6.10.

Table 6.10: Potential sources of marine pollution.

Serial No.	Potential source
1.	Domestic sewage
2.	Sewage sludge
3.	Industrial wastes
4.	Solid wastes
5.	Shipboard wastes
6.	Pesticides and fertilizers
7.	Dredge spoils

8.	Offshore oil exploration and production wastes
9.	Oil spills
10.	Radioactive wastes
11.	Heat: thermal pollution and Power plants
12.	Continental run-off
13.	Antifouling paints
14.	Barges and other metallic structures
15.	Ocean mining
16.	Precipitation and air borne pollutants
17.	Oil from tanker cleaning and deballasting
18.	Weathering of the earth's crust
19.	Volcanic eruptions
20.	Natural submarine oil seeps

Effects of Water Pollution

Water pollution causes several water borne infectious diseases like Cholera, typhoid etc. Plant nutrients and aquacultural wastes contribute NO_3, PO_4 and NH_3 in the aquatic phase which trigger the growth of algae leading to **eutrophication.** The heavy metals discharged from various industries also accumulate in the body tissues of the aquatic organisms and affect their normal physiological processes and growth. The growth of tiger prawns *Penaeus monodon* cultured with the waste waters of the Calcutta city have been reduced to a significant level in comparison to their normal growth (Mitra *et al.*, 1999). The organic wastes not only reduce the dissolved oxygen concentration of the aquatic phase, but also promote the growth of anaerobic bacteria.

This ultimately pose a negative stress on the fish population of the waterbodies. Oil pollution in coastal zone or estuarine area, often affects light penetration and oxygen level in the column water by creating a film on the surface water. This affects the biotic community structure in the oil spilled area. The important adverse effects caused by water pollutants are shown in table 6.11.

Prevention and Control of Water Pollution

Water is a chemical compound that behaves as an excellent solvent. Hence, it is practically impossible to get water in pure state in nature. Modern society utilizes water as principle and stable resource for domestic, industrial and agricultural purposes, power supply, aquaculture and even for recreation by enclosing it in artificial reservoirs. Pollution of water will definitely cuttle these uses of water and will pose a negative impact on mankind. The various ways and technologies to prevent and control water pollution may be discussed under three broad headings: **[a] Removal of Pollutants, [b] Recycling of Wastes,** and **[c] Implementation of Laws**.

[a] Removal of Pollutants

The existing sewage treatment methods (like primary, secondary and tertiary treatments) are very effective in removing selective pollutants from contaminated water.

Primary treatment is a mechanical process which simply removes solids. The relatively larger solid particles are separated by means of metal screens; sands and small stones settle in grit chamber from which the water passes to the sedimentation tank, where the velocity of water is

Table 6.11: Water pollutants: sources, effects and residence times

Pollutant	Major anthropogenic sources	Adverse effects	Residence time
Oxygen demanding organic wastes (biodegradable)	Domestic sewage, industrial wastes, animal waste	Depletion of dissolved oxygen, death of fish, damage to plant life, foul smell	Days to weeks
Pathogens	Domestic sewage, animal waste	Outbreak of water borne diseases	Days to months
Acids	Industrial wastes, acidic deposition	Death of aquatic organisms, solubility of harmful substances increases	Upto years
Salts	Irrigation, industrial wastes	Increase of salinity, death of aquatic organisms, becomes unfit for drinking purpose, domestic use, industrial use and agricultural use.	Variable
Heavy metals,	Industrial wastes,	Toxic to humans, animals	Months to

arsenic etc.	leaching from soils by acidic substances	and aquatic life	years
Plant nutrients (NO_3, PO_4)	Agricultural, domestic and industrial wastes	Eutrophication, fish death, disruption of community structure	Decades
Heat	Cooling water from power plants and other industries	Solubility of oxygen decreases, harmful to aquatic life	Days
Radioactive substances	Mining of radioactive substances, nuclear power generation	Carcinogenic and mutagenic	Days to years
Chlorine and its compounds	Paper and other industries, water disinfection	Fatal to fish possibly carcinogenic	Variable
Oil and grease	Lubricants and solvents etc., petroleum wastes	Potential damage to ecosystem, taste and odour problems for drinking water	Days to years
Pesticides and herbicides	Agriculture, pesticide and herbicide	Toxic to aquatic life, mammals and humans,	Days to years

	manufacturing units	some chemicals are carcinogenic and mutagenic	
Other synthetic organics (Phenols, ethers, chloroform, nitrosamines etc.)	Industrial effluents	Toxic to aquatic life and humans, some chemicals are carcinogenic	Months to years

Source: W. Viessman and M.J. Hammer: Water supply and Pollution control, Harper and Row Publisher, New York, USA, 1985.

drastically reduced and the small particles as sludge. Scum is removed from the upper most layer.

Secondary treatment is essentially a biological process which has the efficiency to remove most of the organic matter. In the **activated sludge process** of secondary treatment, the incoming sewage is mixed with decomposer bacteria and air of oxygen. Thus the complex organic matter through this treatment is broken down into simpler forms. One important drawback of this method is that if any toxic, industrial chemicals enter the sewage, then the decomposer bacteria may be killed and the entire operation may be inhibited.

Tertiary treatment is able to remove virtually all the remaining contaminants. Water leaving conventional secondary treatment still have most of the original phosphates and nitrates, any persistent insecticides and herbicides, disease causing bacteria and viruses, and perhaps a number of industrial, organic compounds. Waste water that is not subjected to tertiary treatment contains the nutrients on which algae thrive.

The various **physico-chemical techniques** used for removal of chemical, biological or radiobiological pollutants are adsorption, electrodialysis, ion exchange and reverse-osmosis. Of the various techniques, the reverse-osmosis technique is based on the removal of salts and other substances from water by forcing the later through a semipermeable membrane under a pressure that exceeds the osmotic pressure so that the flow is in the reverse direction to the normal osmotic flow. In practice, this involves a porous membrane whose chemical nature is such that it has a preferential attraction for solvent while repulsion for the solute. Reverse-osmosis is commonly used to desalinate brackish water and also finds suitable, effective and economical method for the purification of water polluted by sewage effluents.

[b] Recycling of Wastes

Physico-chemical Approach

Various types of wastes like paper pulp, municipal and industrial effluents, sewage etc. can be recycled for beneficial purposes. The urban wastes, for example, can be utilized to generate cheaper fuel gas and electricity.

Electrodialysis is one of the membrane processes developed commercially for desalting saline water to get potable water. This technique can be applied for treatment of wash effluents, especially in electroplating industries (Mayer *et al.,* 1992). The method is used to separate and recover the valuable metals in the form of ions in solution and also enables low ion concentration to be brought upto higher level, which can be reused in the process (Thampy *et al.,* 1997).

Feasibility of chromium recovery and reuse has been studied in details Rajamani *et al.* (1997). The chromium containing waste water including wash water is cleaned and collected in treatment tank. A calculated amount of MgO is added to the liquor and stirred. During this stirring period, the pH gradually rises to the required value of about 8. The stirring process is stopped after the stabilization of pH. The chromium precipitates and settles to a very compact sludge within an hour which is only about 8% by volume of the exhaust chrome liquor. The supernatant liquor is decanted and the sludge is dissolved in H_2SO_4 so that there is the formation of basic chromium sulphate, which can be reused as tanning agent.

This technology can be easily adopted by the Indian tanneries. For a medium scale tannery, with a processing capacity of 5-6 tonnes/day, the capital investment for the chrome recovery system is about 10,50,000. By adopting the tanning procedure with 7-8% basic chromium sulphate and considering the production period of 250 days per year, the annual consumption of basic chromium sulphate will be 120 tonnes. The wastage would be 35-40% of the used chromium and about 35 tonnes of chromium per year can be recovered and reused which is equivalent to Rs. 9,80,000.

Biological Approach

The polluted water can also be reused for fish culture, oyster culture, algal culture and for plant growth. Few instances of biological treatment of polluted water are cited here:

[A] Wetland plants control pollution by absorbing nutrients with a greater efficiency and lower cost, than conventional water treatment facilities. In addition suspended solids are eliminated and clarity of the water is restored that flows through the wetlands. These plants can later be used for methane generation.

Water hyacinth, which was once considered a problematic aquatic weed, has been found to be unique absorber of heavy metals and nutrients. However periodic harvesting of these weeds from the aquatic system is an important step in this technology. Although much slower than conventional sewage treatment facilities, the high initial investment and the energy costs involved in operating conventional treatment plants makes the hyacinth ponds less than half as expensive. In the United States, the city of San Diego, California, is spending $3.5 million in cultivating water hyacinth for waste water treatment. In India, the Central Leather Research Institute has experimentally used water hyacinth to clean tannery effluents which pollute the ground water resources in Madras. Hyacinth has also been used to augment sewage treatment in Calcutta's Salt Lakes swamp.

Recycling of water hyacinth after treating the polluted water has also high feasibility as the weed has multiple benefits. The ash of the plant contains 30% potash and 13% lime, making it an excellent fertilizer. The plant is also useful as animal and fish feeds, and for paper products, and can be converted to biogas.

[B] A recent work done by Niyogi *et al.* (1997) in the Haldia industrial zone of West Bengal, has revealed the excellent growth of some mangrove species in the estuarine water contaminated with heavy metals. Inspite of considerable concentrations of zinc, copper and lead in the aquatic phase and vegetative parts (root, stem and leaves) of the mangrove species *Sonneratia apetala,* the growth rate has been accelerated, proving their excellent absorbing capacity of heavy metals. These plants can later be used as source of wood for fencing, boat manufacturing etc.

[C] The wastes from the shrimp culture farms which mainly consist of ammonia, nitrate and phosphate can also be recycled by culturing edible oyster with the released water from the shrimp farms. Infact due to excessive nutrient concentrations in the released water, an enormous increase of algal population take place in the aquatic phase (algal bloom). However, as these algae are the food for edible oysters, therefore their excess density can be checked by introducing edible oyster species (like *Saccostrea cucullata, Crassostrea cuttackensis, Crassostrea madrasensis* etc.) in the treatment pond of the shrimp culture farms. This technology will, thus help to convert the aquacultural wastes into useful oyster biomass.

[C] Implementation of Laws

In order to control the water pollution, strict implementation of the laws like water (Prevention and Control of Water Pollution) Act, 1974 and the Environmental (Protection) Act, 1986 are extremely important. As a part of protecting and improving the quality of the environment, the standards for emission or discharge of environmental pollutants from the industries, operations or processes has been specified. Any deviation from these specified values must be seriously treated to discourage pollution from point and non-point sources.

The legal actions to ban the semi-intensive shrimp culture farms in the coastal zone of India is a strict step in controlling the environmental pollution due to aquacultural industries. Considering the various negative impacts like salinization of agricultural land (due to saline water intrusion), deterioration of land fertility, release of excessive nutrients and antibiotics in the coastal waters etc., the Supreme Court of India ordered the coastal states on 11th December, 1996 that all aquacultural farms other than the traditional and improved traditional ones, operating within

500 metres of the high tide level (HTL) in the coastal zone be demolished by March, 1997. The Court also directed that no shrimp farm could be set up within 500 metres of HTL in the zone. In accordance with the principles of 'sustainable development' and 'polluter pays', the Court also directed the government to set up a body under clause (3) of Section 3 of the Environment Protection Act, 1986. The said body will issue permits to the farmers of traditional aquaculture to adopt improved traditional farming system. Aqua-farming in mangrove areas, estuaries, wetlands, salt pans and public land is also prohibited. Furthermore, the Court ruled that the owners of the closed aquaculture units would be liable to pay six years wages as compensation to the workers as well as for eco-restoration of the affected areas by way of creating an Environment Protection Fund.

The Court order was expected to affect Rs. 2,000 crore aqua-farming industry and aqua crops of 1.10 lakh ha. of the coastal land in Andhra Pradesh, Tamil Nadu and Kerala (Sen, 1998). Many workers like farm managers, aquaculturists, water analyzer, feed supplier who were directly or indirectly linked with the industry also faced a gloomy situation no doubt, but a comparative study done to analyze the conditions of coastal waters around Sagar Island (in the Hugli estuarine system) after the implementation of the order of the Supreme Court of India revealed the picture of restoration of water quality in terms of nutrients and dissolved oxygen.

6.4. SOIL POLLUTION

Soil is usually polluted with dumping of solid wastes generated in the household and manufacturing units. Soil pollution can be defined as the deterioration of the soil quality and soil fertility due to addition of some foreign elements. Some common soil pollutants are stated here:

Domestic wastes

These include kitchen garbage, broken bottles, cloth rags, ash etc. Plastic waste is an important component of domestic refuse, which being non-biodegradable in nature, is usually burnt. When PVC (polyvinyl chloride) is burnt, it emits highly corrosive hydrochloric acid, which is an extreme nuisance in the operation of refuse incinerators. When an inadequately high temperature is used in burning waste PVC, phosgene is liberated, which is highly poisonous.

The composition of domestic waste is getting changed with the passage of time. Previously there was a time when ash constituted about 60-65% of the waste. However, with the replacement of coal heating by gas, oil etc., this component has been reduced to a great extent since late nineties (only 12-28%). The percentage of paper and plastics have become dominant in the waste which is reflection of the excessive growth of consumer products based industries.

Industrial wastes

These wastes are mainly composed of fly wash, brine mud, metal scraps, dyes, plastics and many wastes of small scale industries. The soil quality in and around the East Calcutta Wetlands has deteriorated to a great extent due to deposition of chromium and lead in the wetland soil. The main sources of these heavy metals are the tannery wastes and the industrial wastes contaminated mainly with lead (Mitra *et al.*, 1999).

Agricultural chemicals and fertilizers

In this category various herbicides, fungicides, pesticides and variety of chemicals which are used for high productivity are included. Broad spectrum pesticides are effective in many situations, but they have serious flaws as well. They destroy populations of predators, upset ecosystems and poison people and animals. Many other pesticides besides DDT have been in use during the past few years, which caused serious soil pollution. Some of the most common pesticides are listed in Table 6.12.

Table 6.12: Lists of common chemical pesticides

Compound Class	Examples	Persistence in the environment
Organochlorides (also called chlorinated hydrocarbons)	Aldrin Chlordane DDD DDT Dieldrin Endosulfan Endrin Heptachlor Kepone Lindane Mirex Toxaphene	High; 5 to 15 years
Organophosphates	Azodrin Diazinon	Intermediate; 1 week to

	Malathion Parathion Phosdrin	several months
Carbamates	Sevin Matacil Temik Zectran Zineb	Low; 2 weeks or less

Acid rain

Acid rain is due to air pollution, which in turn causes soil pollution. Acid rain severely reduces the soil pH and often poses adverse effect on the animal community of the soil. The crop production is also greatly affected due to alteration of the soil pH.

Animal wastes

The dumping of human excreta and wastes from cow sheds and slaughter houses befouls the land. This category of wastes mainly include cattle shed wastes, fisheries wastes, prawn wastes, squilla, frog wastes, carcasses, slaughter house waste, wool wax etc.

Effects of Soil Pollution

Pollution of soil can pose a negative impact on the biotic community not only by affecting the crop production and causing bioaccumulation of toxic substances in the tissues of the producers and consumers, but also by posing a stress on the beneficial bacteria (like nitrogen fixing bacteria) present in the soil. The important adverse effects of soil pollution are listed here in points.

1. Acid deposition in soil can damage trees, crops and other plants. Acid deposition as a result of human activities is already a serious problem in western and central Europe, Scandinavia, north-eastern United States, south-eastern Canada and is expected to become a problem in other areas.

2. Chlorinated hydrocarbons used as pesticides, are persistent and pass into food chain. They are extremely harmful (after biomagnification) within the body tissues of the members of higher trophic level as they cause softening of brain, cerebral haemorrhage, liver damage, thinning of egg shells in birds and disturbance in the release mechanism of sex hormones.

3. Organopesticides (like Malathion, Parathion, Diazonin, Carbamates etc.) are poisonous to workers who handle them as they cause nausea, vomiting, diarrhoea and muscular tremors.

4. Pesticides also cause an adverse effect on the beneficial organisms present in the soil. Besides killing various living organisms present on the surface of the soil, they reach even the deeper layers through tilling and irrigation on the land, killing much more organisms found in the deeper layer of the soil. With the continuous use of pesticides, the soil micro-organisms lose their capacity of nitrogen fixation.

5. Nitrogen fertilizers produce toxic concentrations of nitrite and nitrate in the leaves and fruits of plants. Nitrate containing canned food not only corrodes tin lining of the can, but also produces vitreous oxide. In the alimentary canal, the nitrate of food is converted

into nitrite. Nitrite finally enters in blood, combines with haemoglobin and produces met-haemoglobin. As a result of this, the oxygen carrying capacity of blood is reduced.

6. Cattle grazing on grass where the soil is contaminated with fluoride (released form factories that process phosphate deposits from commercial inorganic fertilizers) suffer from **fluorosis**. It is a condition that reduces the milk production of cattle, attacks their bones and causes lameness, severe emaciation and in extreme cases may lead to death.

7. The industrial wastes (both solid and liquid) are dumped into the soil without any adequate treatment. This deteriorates the quality of soil and affects the vegetables and crops grown in the soil. A survey done during 1990-95 on some vegetables grown at Dhapa (a place at the eastern fringe of Calcutta) reveals the presence of iron, zinc, copper, manganese, lead, cadmium and chromium in some selective vegetables (Table 6.13).

The basic cause of accumulation of heavy metals in different vegetables is the build-up of heavy metals in the soil due to long term indiscriminate application of raw sewage water. But when treated sewage is used for irrigation, the concentrations of metals in the vegetables is comparatively lower. Studies indicate that metal concentration is five times higher in raw sewage irrigated vegetables compared to treated sewage fed vegetables (Mitra, 1996).

Table 6.13: Concentrations of heavy metals in vegetables grown in Eastern fringe of Calcutta (as dry weight basis).

Vegetable	Parts	Fe (%)	Zn (%)	Cu (ppm)	Mn (ppm)	Pb (ppm)	Cd (ppm)	Cr (ppm)
Cauliflower	Head	0.10	0.09	8.0	11.0	70.0	Trace	Trace
	Leaf	0.16	0.11	36.0	21.0	15.0	1.5	0.9
	Stem	0.10	0.09	12.0	8.0	45.0	Trace	Trace
	Root	0.18	0.09	8.0	8.0	60.0	1.0	0.8
Spinach	Leaf	0.28	0.09	40.0	25.0	68.8	Trace	Trace
	Stem	0.26	0.13	24.0	22.0	12.5	Trace	Trace
Radish	Leaf	0.40	0.15	40.0	20.0	30.1	Trace	0.9
	Stem	0.10	0.09	20.0	11.0	20.7	Trace	1.1
	Root	0.08	0.05	6.0	9.0	12.5	1.2	1.2

Source: Gupta *et al.* (1990), Mitra (1993)

Control of Soil Pollution

In recent years various advanced and eco-friendly technologies are used to maintain the quality of the soil which are discussed here:

Modification of cultivation procedure

The traditional and similar type of cultivation that is practiced year after year in any agricultural plot invites pests that feed on a particular crop and a sort of synchronization develops between the pests and the life cycle of crop. These pests develop a crop specificity with the passage of time and start multiplying their population. However, this can be prevented by **crop rotation**. Rotating the type of crop grown in a field from year to year can be used to reduce populations in the soil or crop residues of non-migrating pests that feed on a particular crop. Planting times can also be altered to ensure that most of the pest populations starve to death before the crop is available. These natural processes can minimize the dependency of the agriculturists on pesticides, which greatly deteriorates the natural soil productivity.

Biological control of pests

Agricultural scientists of modern *era* have strong inclination towards biological control for regulating the population of insects, weeds, rodents and other pests by reintroducing effective natural predators, parasites, and pathogens (disease causing bacteria or viruses) or importing new ones. This is mainly because the process is cost-effective and also eco-friendly. Examples of biological control include using lady bugs and praying mantises to control aphids and employing speck-sized parasitic wasps to control various crop-eating moths and flies such as the leaf miner. A bacterial agent (*Bacillus thuringiensis*) sold commercially in dry powdered form, has shown its efficiency in controlling many strains of leaf eating caterpillars, mosquitoes and gypsy moths. Using effective organisms to suppress the pest populations is an important technology to upgrade the soil quality.

Use of pheromones and attractants

Use of various pheromones and hormones to control the insect pest species is an interesting approach to maintain the soil fertility. Pheromones extracted from an insect pest species or synthesized in the laboratory, if spread in an infested area, can attract insects (pests) of opposite sex, which can be finally trapped. Recent researches indicate that instead of using pheromones to trap insect pests, it is more effective to use them to lure natural predators of the target pests into fields and gardens.

Construction of pits

Industrial wastes are hazardous and often biomagnify through food chain. Hence pits or low lying areas are selected for dumping of industrial wastes on which no crops or vegetables are grown. However, plant species having high timber value or used as source of wood for fencing and manufacturing boats may be planted on these pits or adjacent to these pits.

Recycling of wastes

In many cases, the solid wastes dumped on the soil can be put to direct use. Fly ash from coal fired power stations is a major industrial waste product which blanket over the original top soil and affect the agricultural productivity. About 55 tons of coal are needed to generate 1 KW of power and fly ash constitutes 15-20% of the total coal consumption. The different products that can be manufactured from fly ash are cement and concrete, asphalt filters, blocks for civil engineering works etc. In many areas fly ash is also used for filling low lands. The use of fly ash is gaining importance because of the fact that mineral matter in the fly ash has already been treated during the burning of coal. Hence less heat is needed to fire (hardening by applying heat energy) bricks, which may be partly made from fly ash. The advantages of bricks prepared from fly ash are listed below:

- Firing is possible at low temperature. Hence, cost of bricks becomes lower.
- Drying time required by the bricks is less.
- Fly ash is cheaper than clay.
- With less clay content, the landscape takes longer time to get spoiled.

Composting of refuse

The domestic or slaughter house refuses that cause soil pollution are composted to get rid of the pollution effect. On the basis of microbial reactions involved during the process of composting, there are two mechanisms by which composting of refuse take place. They are aerobic and anaerobic.

In **aerobic** decomposition, organic matter in the refuse is utilized (decomposed) by bacteria in the presence of air (oxygen) to produce carbon-di-oxide and other simpler compounds, that are extremely important for plant growth.

Aerobic fermentation is a very common phenomenon in nature and is the main way through which waste products form fields and forests are converted into humus.

In **anaerobic** fermentation, the process of microbial decomposition occurs in the absence of air with the generation of methane. In nature, this process occurs at the bottom of marshy area. In anaerobic fermentation, sulphur is reduced to hydrogen sulphide and not oxidised to sulphuric acid as witnessed in aerobic respiration. Heat energy produced in this case is also much less in comparison to aerobic fermentation.

An important criteria for carrying out the operation of composting is the C:N ratio of the refuse. If this ratio is much above 35, the refuse materials undergo composting at a slow rate. The optimum C:N ratio is around 27 and below this ratio, the speed of microbial fermentation is retarded because there is inadequate carbon present to provide energy for chemical changes induced in the nitrogen containing compounds. It is for this reason, refuses or substances that are dumped into the soil (like waste plant products, saw dusts etc.) are again blended with substances that have very high C:N ratio (like human or animal excreta, slaughter house remains, fish scraps etc.). The average composition of some ideal blending substances are highlighted in table 6.14.

Table 6.14: Average composition of animal excreta, slaughter house and farm wastes.

Parameters	Human or animal excreta	Slaughter house waste	Farm waste
Moisture content (%)	66-68	75-80	10-60
Organic matter (% on dry weight basis)	88-97	80-95	25-35
Nitrogen (% on dry weight basis)	5-7	8-11	0.4-0.8
Phosphorus as P_2O_5 (% on dry weight basis)	3-5	3-3.5	0.2-0.5
Potassium as K_2O (% on dry weight basis)	1-2.5	2-2.5	0.8-1.5
Carbon (% on dry weight basis)	40-55	14-17	12-17
Calcium as CaO (% on dry weight basis)	4-5	3-3.5	4-7.5

The process of composting is a unique technology to upgrade soil quality as the harmful pathogens and parasites are destroyed and heat killed during the operation (since a temperature of 50-70°C is maintained) and the processed end products can be used to increase soil fertility. The average composition of an ideal compost is shown in Table 6.15.

Table 6.15: Average composition of an ideal compost

Parameters	Composition
Organic matter	25-50%
Carbon	8-40%
Nitrogen	0.5-3.5%
Phosphorus as P_2O_5	0.5-3.5%
K as K_2O	0.5-2%
Ca as CaO	2-7%
Incombustible matter	20-60%

6.4. CONCEPT OF INDICATOR SPECIES

The environment, in which we live is highly dynamic with several environmental factors constantly oscillating with time. Sometimes it happens that a particular variable which was dominant in the dawn becomes totally supressed in the afternoon leaving no trace of its existence in the system. With the ingression of tides, the incoming sea water increases the salinity of the estuarine system where there was almost fresh water during the period of ebb. Thus change of environmental variables with respect to time is a game of nature where many variables are difficult to identify at a particular instant of time. Infact, the ecologist constantly employs organisms as indicators in exploring new situation or analysing the fluctuation of environmental variables.

Use of selective species of plants as indicators of mineral substances was pointed out by many workers (Cannon, 1952, 1953, and 1954). When deep rooted plants like pines and junipers grow over uranium deposits, the vegetative parts of the plant contain a high concentration of uranium. These plant parts on ashing can exhibit the concentration of uranium if analysed fluorimetrically. A concentration of 2 ppm uranium or more in the ash is considered indicative of commercially usable deposits underground. Since selenium is often associated with uranium ore, selenium indicating plants like *Astragalus* sp. may be useful in locating deposits in the rocky mountain region. Similarly, where sulphur and uranium are associated, the sulphur accumulating members of the mustard and lily family provide useful indicators.

Today the concept of "**indicator species**" or "**indicator organisms**" have become an important tool in monitoring an environmental system. The sudden inclination towards this concept is primarily due to the following two reasons:

1. The environmental variables are in a state of constant oscillation and hence chemical analysis by collecting sample at a particular instant of time cannot give a correct picture in many cases. A study done in 1998 by Choudhury *et al.* in Haldia industrial belt reveals a concentration of dissolved lead 22 μgl^{-1} during the period of low tide, but water of the same sampling point when analysed during high tide exhibited a concentration of 13.6 μgl^{-1} dissolved lead. Although it may be an effect of precipitation due to high salinity, but getting a correct picture of the environment by analysing the water sample at a particular instant of time, becomes almost impossible.

2. The cost of chemical analysis of environmental variables sometime becomes very high, due to which a large number of samples can not be accommodated for analysis within the budget. Moreover the intercalibration exercise is a vital part of chemical analysis to maintain the quality of the data, the accessibility of which poses a trouble to the users of the under developed countries.

It is in this context, the fluctuation of population of a particular species or a group of species can be used to understand the change in environment. The two important properties serving as the foundation stone to apply the concept of indicator species for analysing an environment are: (1) population fluctuation and (2) bioaccumulation.

When a system becomes unfavourable, the sensitive species starts eliminating from the area and the opportunistic organisms start to flourish. Under such circumstances, the ecological indices (like Shannon Weaner index, index of dominance, evenness index, richness index etc.) show a change in their values e.g., the index of dominance increases for ciliated protozoa if a system is exposed to organic pollution. This is because ciliated protozoa is primarily bacteria feeders whose population flourishes due to increased amount of organic load. Under such circumstances, the values of Shannon Weaner species diversity index, richness index and evenness index become low. Ecological responses of flora and fauna has been used extensively during the last twenty years (Mackenthum, 1966; Hawkes, 1977; Sarkar *et al.,* 1985). There has been a great thrust for developing water quality criteria based on quantitative enumeration of ecological data which are called biological indices of water quality. Several systems exist today to judge the water quality depending on the biological indicators. The fundamental concept underlying all these systems is basically the same, i.e., the biota respond to pollution and the type of response is regulated by the amount of pollution and the time of exposure.

Three major groups of responses that are quantified for biological monitoring of waters are: (1) ecological responses (2) physiological responses and (3) toxicological responses. These categories again encompass several indices. The approach that are adopted and the group of organisms selected for data acquisition and interpretation differ in each method as indicated below:

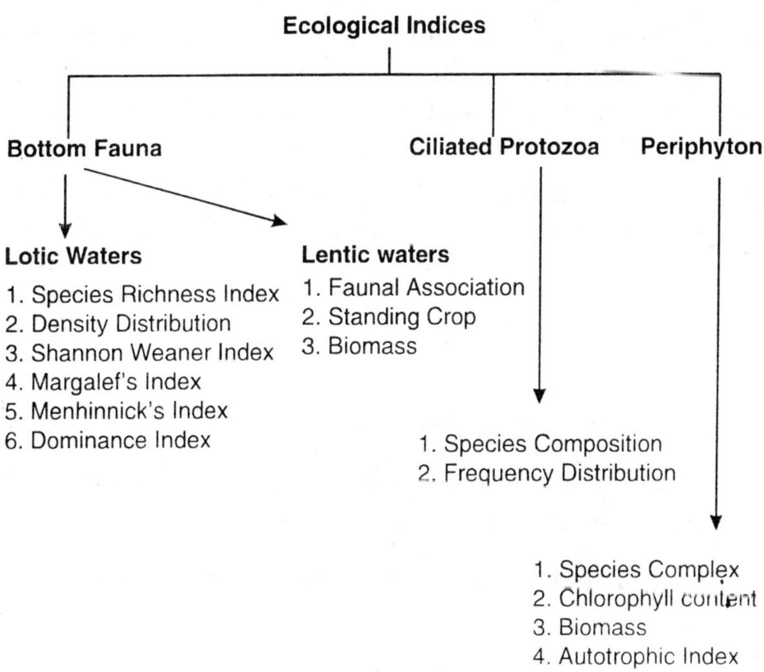

The concentrations of conservative wastes in the body tissues of biota can also act as pathfinder of water quality assessment. In many cases the tissue deposition of heavy metals or pesticides synchronizes with their concentrations in the ambient media (sediment and water). The statistical significance (usually computed through r-value) of tissue concentrations of pollutants with that of the ambient media is a criteria of identifying indicator species of a particular pollutant or a group of pollutants. The famous '**mussel watch**' concept has been evolved as a programme for monitoring global marine pollution for metals, pesticides, petroleum hydrocarbons and radionuclides. The concept takes into account the well known ability of marine molluscs like mussels to concentrate pollution to very high levels as compared to the levels in seawater, thus acting as '**sentinels**'. In practical field, it is observed that many of the organisms either shift from the reference site due to fluctuation of environmental variables or die completely with the change of salinity or pH leaving only the dead shells. Hence, some important requisites were fixed for being included in the domain of indicator species which are listed below:

- The organisms must have wide range of tolerance to environmental variables.
- The organisms must be readily available and accessible.
- The organisms should be relatively stationary in most parts of the life cycle.
- The organisms should be available at all times of the year.
- The organisms should have wide range of distribution.

For estuarine purposes, the organisms should be sufficiently tolerant to low salinities and high suspended to solids to penetrate a reasonable distance upstream (Phillips 1977 and 1980).

References

1. Baker, H. G. & Hurd, P. D. (1968). Intrafloral ecology. Annu. Rev. Entomol. 13, 385-415.
2. Banerjee, R. (1985). Captive Rearing of olive Ridley Turtle. Hamdryad., 12-14.
3. Blasco, F. (1977). Outlines of ecology, Botany & Forestry of the Mangals of the Indian Subcontinent, West Coastal Ecosystem. Elsevier Scientific Publishing Company, Oxford.
4. Cairns, J. Jr & Dickson, K. L. (1971). A simple method for the biological assesement of the effects of waste discharges on aquatic bottom-dwelling organisms. J. Wat. Pollut. Control Fed. 43, 755-772.
5. Chatterji, A. & Abidi, S.A.H. (1993). The indian horseshoe crab – a living fossil. Journal of Indian Ocean Studies, 1, No. 1, 43-48.
6. Chattopadhyay, R. R., Sarkar, S. K., Ganguly, S. & Basu, T. K. (1993). Biochemical and toxicity studies with Azadirachta indica leaf extract. J. Ecobiol. 5(3). 233-235.
7. Chaudhuri, A. B. & Choudhury, A. (1994). Mangroves of the Sunderbans, India, IUCN.
8. Chew, R. M. & Chew, A. E. (1970). Energy relationships of the mammals of a desert shrub (Larrea tridentata)
9. Chopra, R. N., Nayar, S. L. & Chopra, I.C. (1956).

Azadirachta indica In: Glossary of Indian Medical Plants, CSIR.

10. Das, D., Sikdar, K. & Chatterjee, A.K. (1994). Potential of *Azolla pinnata* as Biogas Generator and as a Fish-Feed. *Indian J. Environ. HLTH.* **36(3),** 186-191.
11. Elton, Charles. (1927). Animal Ecology. Mac Millan Company, New York. (2nd ed. 1935; 3rd ed. 1947).
12. Evans, F.C. (1956). Ecosystem as the basic unit in ecology. *Science,* **123**, 1127-1128.
13. Gates, D.M. (1965). Heat, radiant and sensible. Meteorological Monographs, **6**, 1-26.
14. Gleason, H.A. (1926). The individualistic concept of the plant association. *Bull. Torrey Bot. Club,* **53**, 1-20.
15. Government of India (1992 a). Conservation of Wetlands in India. Ministry of Environment and Forests, New Delhi.
16. Government of India (1992 b). National Conservation Strategy and Policy Statement on Environment and Development. Ministry of Environment and Forests, New Delhi.
17. Grinnell, Joseph. (1928). Presence and absence of animals. *Univ. Calif. Chron.,* **30**, 429-450.
18. IUCN (1989). Marine Protected Areas Needs in the South Asian Seas Region, **2**, India.
19. Kochsiek, K.A., Wilhm, J.L. & Morison, R. (1971). Species diversity of net zooplankton and physicochemical conditions in keystone reservoir, Oklahoma. *Ecology,* **52**, 1119-1125.
20. Kozlovsky, D.G. (1968). A critical evaluation of the trophic level concept. I. Ecological efficiencies. *Ecology,* **49**, 49-60.

21. Kwasniewska, K. (1985). Biodegradation of crystal violet by red oxidative yeasts. *Bull. Environ. Contam. Toxicol.*, **34**, 323-333.
22. Margalef, R. (1968). Perspectives in Ecological Theory. University of Chicago Press, Chicago.
23. Miller, R.S. (1967). Pattern and process in competition. *Advances Ecol. Res.*, **4**, 1-74.
24. Niyogi, S., Basu, S., Mitra, A. & Choudhury, A. (1997). Trace metal concentration in vegetative parts of *Sonneratia apetala.* Indian journal of Environmental Health, NEERI, **39(1),** 69-72.
25. Odum, H.T. & Eugene, P. (1966). Regenerative systems: Discussion (A.H. Brown, discussion leader). In: Human Ecology in Space Flight (D.H. Calloway, ed.). The N.Y. Acad. Sci., pp. 82-119. Interdisciplinary Communications Program, New York.
26. Odum, H.T. (1971). "Environment, Power and Society," Wiley, New York.
27. O'Neill, R.V. (1968). Population energetics of the millipede, *Narceus americanus. Ecology,* **49**, 803-809.
28. Osborne, A., Wanielista, P. & Yousef, A. (1976). Benthic fauna species diversity in six Central Florida lakes in summer. *Hydrobiologia,* **48**, 125-129.
29. Park, T. (1954). Experimental studies of interspecific competition. II. Temperature, humidity and competition in two species of *Tribolium. Physiol. Zool.,* **27**, 177-238.
30. Petrides, G.A. & Swank, W.G. (1966). Estimating the productivity and energy relations of an African elephant population, pp. 831-842 in Proc. 9th. Int. Grassland Cong., Sao Paulo.

31. Phillips, D. J. H. (1977). The use of biological indicator organisms to monitor trace metal pollution in marine and estuarine environments. *A review. Environ. Pollut.* **13**, 281-317.

32. Phillips, D. J. H. (1980). Quantitative aquatic biological indicators. Applied Science Publishers Barking.

33. Phillipson, J. (1966). Ecological energetics. Arnold, London.

34. Reifsynder, W. E. & Lull, H. W. (1965). Radiant energy in relation to forest. *Tech. Bull. No.* 1344, pp. 111, U.S. Dept. Agriculture, Forest Service.

35. Ryther, J.H. (1969). Photosynthesis and fish production in the sea. *Science,* **166**, 72-76.

36. Sani, R.K. & Banerjee, U.C. (1999). Screening for organisms applicable to the decolorization of triphenylmethane dyes and optimization of biotransformation conditions in stirred tank reactor. *Indian J. Env. & Ecoplan,* **2(1)**, 1-9.

37. Schoener, T. (1973). Population growth regulated by intra-specific competition for energy or time: some simple representations. *Theor. Popul. Biol.* **4**, 56-84.

38. Schoener, T. (1974). Competition and the form of habitat shift. *Theor. Popul. Biol.,* **6**, 265-307.

39. Sidhu, S.S. (1963). Studies on Mangrove. *Proc. National Academy of Sciences,* India, **33(b)**, Part 1, 129-136.

40. Simpson, E.H. (1949). Measurement of diversity. *Nature,* 163, 688.

41. UNEP. (1985). Environmental problems of the Marine & Coastal areas of India. UNEP Regional Seas Reports & Studies. No. 59.

42. Warren, C.E., Wales, J.H., Davis, G.E. & Doudoroff, P. (1964). Trout production in an experimental stream enriched with sucrose. J. Wildl. Magmt. 28: 617-660.
43. Whittaker, R.H. (1960). Vegetation of the SisKiyou mountains, Oregon and California. Ecol. Monogr. 30, 279-338.
44. Whittaker, R.H. & Niering, W.A. (1965). Vegetation of the Santa Catalina Mountains, Arizona: a gradient analysis of the south slope. Ecology, 46: 429-452.
45. Wilhm, J.L. and Dorris, T.C. (1968). Biological parameters of water quality. Bio Science, 18: 477-481.
46. Yatome, C., Ogawa, T., Koza, D. & Idaka, E. (1981). Biodegradation of azo and triphenylmethane dyes by Pseudomonas pseudomallei 13 NA J.S.D.C. 97: 166-169.